Aspects of Kolmogorov Complexity

RIVER PUBLISHERS SERIES IN INFORMATION SCIENCE AND TECHNOLOGY

Volume 4

Consulting Series Editor

KWANG-CHENG CHEN
National Taiwan University
Taiwan

Information science and technology enables 21st century into an Internet and multimedia era. Multimedia means the theory and application of filtering, coding, estimating, analyzing, detecting and recognizing, synthesizing, classifying, recording, and reproducing signals by digital and/or analog devices or techniques, while the scope of "signal" includes audio, video, speech, image, musical, multimedia, data/content, geophysical, sonar/radar, bio/medical, sensation, etc. Networking suggests transportation of such multimedia contents among nodes in communication and/or computer networks, to facilitate the ultimate Internet. Theory, technologies, protocols and standards, applications/ services, practice and implementation of wired/wireless networking are all within the scope of this series. We further extend the scope for 21st century life through the knowledge in robotics, machine learning, cognitive science, pattern recognition, quantum/biological/molecular computation and information processing, and applications to health and society advance.

- Communication/Computer Networking Technologies and Applications
- Queuing Theory, Optimization, Operation Research, Statistical Theory and Applications
- Multimedia/Speech/Video Processing, Theory and Applications of Signal Processing
- Computation and Information Processing, Machine Intelligence, Cognitive Science, and Decision

For a list of other books in this series, see final page.

Aspects of Kolmogorov Complexity

The Physics of Information

Bradley S. Tice

Advanced Human Design
Cupertino, California, U.S.A.

River Publishers

Aalborg

ISBN 978-87-92329-26-4 (hardback)

Published, sold and distributed by:
River Publishers
P.O. Box 1657
Algade 42
9000 Aalborg
Denmark

Tel.: +45369953197
www.riverpublishers.com

I think that these observations should serve as an antidote to the excessive egotism, competitiveness and the foolish fights over priority that poison science. No Scientific idea has only one name on it; they are the Joint production of the best minds in the human race, Building on each other's insights over the course of History.*

*Concluding comments by Gregory J. Chaitin in his new book *Meta Math* (Pantheon Books, New York, 2005), page 142.

In Memoriam

Lilburn Trent Tice, Junior*
(October 10, 1927 to February 14, 2008)

*My father was a graduate of The University of Kansas, mathematics (almost chemical engineering), and had an ABD, 'All But Dissertation', in the graduate school, mathematics and computer science, at Stanford University. He worked a decade for Lockheed Aerospace and retired as a mathematician from the Stanford Linear Accelerator Center, S.L.A.C., to enjoy his free time with his Peet's Coffee and his granddaughter Rachel. A collection of his university and professional books is being given as a gift in his name to the library at the University of California at Merced.

Table of Contents

Acknowledgement

I would like to thank my mother, Paula Nanette Tice, for being the primary external reader and editor for this academic work. She has decades of experience in the electronics industry that started with General Electric in the 1950s in their computer lab's division.[1] I would also like to thank Advanced Human Design for the research grant for the time and materials to study for this thesis.

[1] A history of General Electric's interest into computers is summed up in Homer R. Oldfield's *King of the Seven Dwarfs* (Los Alamitos, California: IEEE Computer Society Press, 1996).

Note: My father, Lilburn T. Tice Jr., has had extensive professional experience with the field of computation and is a retired mathematician from the Stanford Linear Accelerator Center, S.L.A.C., located in Menlo Park, California U.S.A.

Preface

This thesis started with a clear and defined idea of the topic, and through the process of development has gone from a collection of research notes and papers, to a frenzy of reading, revisions and historical and literary review. Not unlike a Chinese fire drill. The examination of the lengths of binary bits, like the question of the length of Lady Godiva's hair, comes down to the very long and the very short. In other words, a question of the infinite and finite. Within these two quantities lies another group of qualities that will be examined in this thesis: randomness and non-randomness as they relate to a series of binary bits, strings, that make up the focus of Algorithmic Information Theory and Kolmogorov Complexity. The research component of this thesis is in addressing a new measure for the standard of randomness in a binary bit string. This 'sub-maximal measure of Kolmogorov Complexity' redefines the current notion of the measure of randomness in a binary bit string.

I have included photographic plates of artwork I find pertinent to this subject, as I have chosen the prints of M.C. Eischer and a sculpture by Auguste Rodin for both esthetic and intellectual merits. Art and science are not unmutual and I find the art compliments the ideas found in this work.[1]

---

[1] Note: A very odd, even bizarre, little book on art and science that involves communication theory and Shannon's information theory is John R. Pierce's *Science, Art, and Communication* (Clarkson N. Potter, New York, 1968).

Second Preface

The first preface was done at the conclusion of 2006 when I had originally 'finished' the thesis, or rather the thesis had finished me, and 2007 was used to develop the ternary, quaternary, and quincinary based systems for communications applications. A poster on the quincinary based system was delivered in January of 2008 and I began to revise this thesis in March of 2008. What marked my attention to this topic was my father's interest in the radix 5 based poster delivered in January 2008 as he thought it was a 'clever' idea and made contributions to the understanding of statistical communication theory at a level that approached a novelty, his word for a seminal concept, and felt it was on a standard of physics not unlike what was awarded the Nobel Prize in Physics for 2007. Unfortunately, my father passed away on February 14, 2008, leaving me with the motivation to upgrade this thesis and publish my ideas on the subject.

Third Preface

I am pleased that a third preface is needed to address this revised edition of my dissertation as a published book by River Publishers, Denmark in 2009. I have removed the post-script section found in the original dissertation and I have high-lighted the important terms found in the body of the text of the dissertation to make it more perspicuous for referencing.

An index has been added to aid in searching for names and content terminology. I realize that the concepts found in this dissertation are fundamental to statistics; a formalized notion of randomness, physics; a new measure for Kolmogorov complexity, computer science; the symbolic space multiplier program, and mathematics; statistics and probability theory.

I am grateful that my publisher has found merit in publishing this monograph as a revised edition to the original dissertation. It reflects well on the type of publishing firm that would take a chance on publishing a novel idea in the competitive world of technical and scientific publishing.

I hope that my work in this area will lead other researchers to develop and expand on my ideas in the future and continue to push the limits of knowledge as we know it. Only by pushing these known limits of thought can we discover the new and add to the ever growing tree of knowledge.

List of Plates

The following is a list of plates used in this thesis. The work is by M.C. Eischer and it is acknowledged that by permission by Cordon House that they are used in this work.

Plate 1   Hand with Reflecting Sphere

Plate 2   Three Spheres I

Plate 3   Three Spheres II

Plate 4   Gravitation

Plate 5   Relativity

Plate 6   Sphere Surface with Fish

Plate 7   Sphere Spirals

Plate 8   Morbius Strip II

Plate 9[1]   The Thinker

---

[1] Note: The Thinker by Auguste Rodin. By special permission to use by the Stanford University Museum of Art 1988.106. Gift of Gerald Cantor Art Foundation. Photo Credit Frank Wing Photography.

Plate 1

1

Introduction

The thesis addresses the history of *Algorithmic Information Theory*, also known as *Kolmogorov Complexity*, with the research section adding to this history with the examination of a '*sub-maximal measure of Kolmogorov Complexity*'. I will use the terms Algorithmic Information Theory and Kolmogorov Complexity synonymously during the course of this thesis.

Landauer has stated that information has a physical form (Landauer, 1984: 161 and 1993). Landauer notes that the study of the ultimate limits in computing is in its early stages and it is easier to ask questions than to answer them (Landauer, 1984: 161).

The thesis is in some respects two divergent works in that the review of literature is more a historical time line told as a story and the remaining chapters the mechanics of Kolmogorov complexity. While not fiction, I am in agreement with Levin (2006) that 'science without storytelling collapses into a set of equations or a ledger full of data' (Levin, 2006: 45).

The index of terminology should help in defining major ideas found in this thesis and the chapter on the history of *algorithmic information theory* gives a short overview of its founding in the mid-1960s to the present. The research aspect to this work is in the chapter on a *sub-maximal measure of Kolmogorov Complexity* that presents a new measure of the randomness of a binary bit string. The section on binary, ternary and quaternary-based systems of symbols is followed by a chapter on monochromatic and chromatic symbols as they relate to writing, reading, and printing entropy and information. The last chapter is on data compression for algorithmic information theory and how this compression relates to fundamental aspects of information theory.

In working on this thesis I am reminded of the comment by the Hungarian biochemist Albert Szentz-Gyorgyi (1893–1986) that scientific research involves seeing what everyone else has seen but thinking what no one else has thought (Atkins, 2005: 113).

Plate 2

2

Index of Terminology

Algorithm

The Laicization of the famous Islamic mathematician al-Khuwarizmi (783–850) whose name translated to became Algorithm (Ifrah, 2000: 531). Originally the name for the Indian system of a zero with nine digits, Algorism, and the methods used by that system for calculation, it would ultimately acquire the current usage, Algorithm, in the fields of applied and theoretical computing (Ifrah, 2000: 531; Ifrah, 2001: 74). In this thesis, an algorithm is a text that gives instructions on how to proceed from inputs to the result (Jones, 1997: 12).

Algorithmic Information Theory

Also known as Kolmogorov Complexity (see Endnote #1). Kolmogorov indented in developing Algorithmic complexity as a measure of the information content of individual objects (Li and Vitanyi, 1993/1997: 65). Shannon (1948) considered information an 'ensemble' were as Kolmogorov was only interested in 'countable' ensembles (Li and Vitanyi, 1993/1997: 65). Kolmogorov complexity is the a string, a length of binary bits, that contains 'regularities' that allows it to have a shorter description than itself (Li and Vitanyi, 1993/1997: v). In other words, the description of that length of binary bits is compressible by the non-random sequence of binary bits or strings (Li and Vitanyi, 1993/1997: v).

Alphabet

A fixed set of symbols in formal theory used to form strings. These symbols may use the Roman alphabet, Arabic digits or a binary alphabet (Du and Ko, 2001: 1). A string of a binary alphabet is known as a binary string (Du and Ko, 2001: 1),

Codes

Are procedures to represent data to be compressed based on probabilities provided by a model.

Compression

A process of reducing the amount of information bearing units, usually bits, from the original size without the loss of the original information content.

Data

Data refers to the coded representations of numbers, alphabetic characters, and special characters that are used to perform operations of computation (Bitter, 1992: 223). Data are the units of 'reality', facts, perceived aspects of the physical world, that when taken in context, become a form of information, a unified meaning of the data. This thesis will examine data within the formal context of information theory and Kolmogorov Complexity (Endnote #2).

Digit

A single symbol or character representing an integral quantity (Richards, 1955: 5).

Entropy

Algorithmic entropy rate measures how random a process is when viewed as a computer (Crutchfield, 2003: 38).

Functions

A mathematical function is a set (Jones, 1997: 12). A function associates a result with each input but does not say 'how' such a result can be computed (Jones, 1997: 12).

Information

Information is dimensionless and represents pure numbers (Brillouin, 1956/1962: 3). It is represented by two letters or symbols, a [0] and a [1] of which have no semantic meaning other than being the opposite of the other symbol. Information is a form of entropy in that a certain expression can be defined to measure the amount of information in a given operation (Brillouin, 1956/1962: 1).

Information and Meaning

The type of information addressed in this thesis is not the standard notion of information, such as found in natural languages and the study of 'semantics', but rather a precise and defined meaning the deals with the engineering concept of signal content as a symbolic representation by way of a binary set of notations, in this case being [1] and [0]. Natural languages have inherent ambiguity that evades even the most accomplished 'cartographers' of the structure of language. As an example, Chomsky uses as a 'nonsensical' sentence, but grammatically correct to a native speaker of the English language, the following sentence: 'Colorless green ideas sleep furiously' (Chomsky, 1957: 15). Chomsky is being linguistically naíve in that a poet, perhaps a 'beat' poet, may consider it to have meaning well beyond the 'normal' mundane 'everyday' use of speech (Endnote #3). This is the main reason why information theory, and Kolmogorov Complexity, are assigned specific values that are binary in nature and only have the 'meaning' of being the opposite of the other, either a [0] or a [1].

Length

The length is the number of symbols contained in a string, hence the length of a string (Du and Ko, 2001: 2).

A Turing Machine

A Turing machine is an abstract device, which performs a sequence of elementary operations in a succession of discrete stages (Ifrah, 2001: 275). The device has the following features (Ifrah, 2001: 275–276):

1. A finite repertoire of symbols.
2. A potentially infinite tape.
3. An erasing device.
4. A reading and writing device.
5. A displacement device.
6. A situation table.
7. A control unit.

At each step of the process, the Turing machine will perform the following (Ifrah, 2001: 276):

1. The machine is controlled by current state of the control and what symbol, in the square the reading and writing device is situated.
2. Control maybe in either the initial state or change its state.
3. The machine may remain in the same position move, either forwards or backwards, one square.
4. The machine may read, write or erase a symbol in the current square, either replacing it with a different symbol or leaving the square unchanged (Ifrah, 2001: 276).

Non-random

A word is nonrandom if any description of its creation is less large than at its full representation bit by bit.

Numbers

A number is a quantity represented by a group of digits (Richards, 1955: 5).

Quaternary

A radix 4 based character system composed of four separate symbols.

Radix

The number of digit symbols employed in a system (Richards, 1955: 4).

Random

Knuth (1981) considers random sequences generated in a deterministic way to be 'pseudorandom' or 'quasirandom' in that they only 'appear to be random' (Knuth, 1981: 4). The definition used in this thesis is as follows. A word is random if any description of its creation is at least as large as its full representation bit by bit (Hromkovic, 2004: 44).

Sequence

Sequences are usually infinite (Li and Vitanyi, 1997: 12).

Strings

A string is a finite sequence of symbols (Du and Ko, 2001: 1).

Symbol

The symbol 0 and 1 are elements of a character and are not to be confused with the numbers zero and one (Casti, 1996: 140).

Ternary

A base 3 numerical system. Also known as a radix 3 numerical system (Richards, 1955: 3–4).

Trits

A trinary digit that is a base 3 numeral system. Also known as a ternary system (Wikipedia, 'Ternary Numeral System', 2006: 1).

Words

Strings are called words. Relations between strings form a theory called word theory (Du and Ko, 2001: 2).

Plate 3

3

Review of the Literature

The primary area of citations will come from *Algorithmic Information Theories'* founders: R.J. Solomonoff, A.N. Kolmogorov, and G.J. Chaitin. The most numorious citations have come from G.J. Chaitin. The most authoritative text on Algorithmic Information Theory is from M. Li and P. Vitanyi (1993/1997). I have taken the trouble to research timetables of materials cited beyond mere publication dates because it can be years between a papers submission date and publication date and that many publications are small in-house company publications that can take years to get a proper referencing timeline. I will give two examples. R.J. 'Ray' Solomonoff is credited with being the first inventor of Algorithmic Information Theory, also known as *Kolmogorov Complexity*, with his publication in 1964 of a paper giving rise to the concept of this theory.

It was not until years later that Solomonoff's 1960 technical report for the Zator Company in Cambridge, Massachusetts (Technical Report ZTB-138) was widely referenced as a 'seminal' paper in this field, only M. Minsky referred to this early work, in January 1961, in the Proceedings of the I.R.E. (Li and Vitanyi, 1993/1997: 89–90). Minsky cites Solomonoff's 1964 paper in his *Computation: Finite and Infinite Machines* (1967) as a 'great philosophical importance' for inductive inference (Minsky, 1967: 66). Both Chaitin and Kolmogorov were unaware of Solomonoff's 1960 paper as well (Li and Vitanyi, 1993/1997: 92). A.N. Kolmogorov citing him only after 1968 (Li and Vitanyi, 1993/1997: 90). The information about the existence of this paper was taken from the excellent account of the early history of Algorithmic Information Theory by Li and Vitanyi (1993/1997: 89–90).

Almost all the 'general' histories of Algorithmic Information Theory/Kolmogorov Complexity leave out Solomonoff's 1960 paper and it is generally considered to be the mid-1960s that these theories were 'developed' and published by Solomonoff, Komogorov, and Chaitin. There has been a considerable amount of time spent 'cross-checking' even 'primary' sources of information in this field as a 'selective memory' seems to be a common denominator in presenting the priority of the 'who, what, and when' of Algorithmic Information Theory (Endnote #4).

In 1965, when Chaitin was an 18 year old student at the City College of the C.U.N.Y. system, he submitted two papers to the *Journal of the ACM* of which one was published in 1966 as the 'complexity of algorithms following Shannon's coding concepts' and the other, published in 1969 that puts forward the idea of 'Kolmogorov complexity' (Li and Vitanyi, 1993/1997: 92). In other words, the submission times of both papers occurred around the mid-1960 as did Solomonoff's (1964) and Kolmogorov's (1965) papers. The date of publication does not always accurately reflect the date of an idea or concept, let alone the date of the submission of the papers of that idea. Chaitin notes in this book *The Unknowable* (1999) that it was 'unfortunate' that the editor delayed publication of the second paper, they were originally submitted as one paper, and that the referee, a Donald Loveland, 'immediately' sent the entire uncut original paper to Kolmogorov in Moscow (Chaitin, 1999: 85).

Chaitin would be awarded the Belden Mathematical prize and the Nehemiah Gitelson award while a student at City College (Chaitin, 2005: 34). Chaitin was in his second year at City College, 1965, when he was excused by the Dean to prepare his paper, 'papers', for publication, ultimately in the *ACM Journal* of 1966 and 1969 (Chaitin, 2005: 121–122). Chaitin's first publication was while he was a student at the Bronx High school of Science 'An improvement on a theorem of E.F. Moore' in the *IEEE Transactions on Electronic Computers* (EC-14) (1965), pages 466–467 (Chaitin, 1999: 85) and (Li and Vitanyi, 1993/1997: 92). Chaitin was 'self-educated' and does not posses a college degree (Chaitin, 2002: 66).

Some noticeable traits or characteristics were discovered when compiling this review of literature. The first is that Algorithmic Information Theory, or AIT, is also known as Kolmogorov Complexity and depending on the 'bias' of the individual writing on this topic will term it as either, and sometimes as both, names. M. Gell-Mann, the Noble Prize winning physicist, has termed it Algorithmic Information Content, or AIC, but this just refers to the number of bits needed to store a computer program (Gell-Mann, 1994: 38–39; Chaitin, 2006: 76) and (Endnote #5). I have found that Chaitin will use Algorithmic

Information Theory while others, usually those from the Continent, will use Kolmogorov Complexity. *The New Encyclopedia Britannica* (2005) lists Algorithmic Information Theory within the entry for Information Theory (Ge, 2005: 637).

Perhaps it is a nationalistic preference in such naming as I have noticed that some papers, such as Uspensky (1992), that disregard Chaitin's contribution to Algorithmic Information Theory altogether while citing Kolmogorov as the sole inventor of the process. Uspensky even down plays Ray Solomonoff's contribution to the theory in his early paper of 1964 that was, according to Uspensky, unknown to Kolmogorov, who published his results in 1965. Uspensky even states that Solomonoff's 1964 paper 'presented some similar ideas – but in a vague and rather non-mathematical manner' (Uspensky, 1992: 96). Uspensky does not elaborate what is 'vague' or is 'non-mathematical' about Solomonoff's paper but given the fact that Uspensky has stated that the purpose of Kolmogorov's 1965 paper was to bring the notion of complexity to the foundations of information theory, places this comparision in some illustrious company as it was Claude Shannon's ground breaking work *The Mathematical Theory of Communication* in 1948 that gave rise to information theory that was considered by many of Shannon's peers to be not mathematical at all (Shannon and Weaver, 1949; Gilbert, 1966: 320; Doob, 1949, 1959).

Chaitin also states that Solomonoff's papers, published in *Information & Control*, in two parts, that Solomonoff's 'math isn't very good and he doesn't really succeed in doing much with these ideas' (Chaitin, 1999: 86). Chaitin does not mention Solomonoff's earlier work (1960) in this book. Chaitin does mention Solomonoff's 1960 paper in his historical introduction to the 'Algorithmic Information Theory' paper published in the *Journal of IBM Research and Development* (1977) and claims it as being 'the first ideas on algorithmic information theory' that were cited from Minsky's paper of 1961 (Chaitin, 1977: 350; Chaitin, 1987: 38). Chaitin also mentions himself, Kolmogorov and Martin-Lof, along with Solomonoff, as independent contributors to the development of algorithmic information theory (Chaitin, 1987: 39). Chaitin does not mention Kolmogorov's 1968 paper on the subject until 1970 in the 'amended' section of the references to a paper dated that year (Chaitin, 1987: 22).

In the *Encyclopedia of Mathematics* (1988) Barzdin describes an entry to Algorithmic Information Theory that cites Kolmogorov, and Martin-Lof, but fails to give credit to Solomonoff and Chaitin (Barzdin, 1988: 140–142). In a following 'editorial comment' to the Barzdin entry an editor describes the history of algorithmic information theory as being 'originated in the independent work' of Solomonoff, Kolomogorov and Chaitin (Barzdin, 1988: 142). This divide does not have the rancor that one would find in the Behavior verses Innate debate found in the fields of psychology and linguistics, especially the Skinner (behavorist school) verses the Chomsky (cognitive school) debate, but there is a line of demarkation.

My initial introduction into the field of Algorithmic Information Theory was Chaitin's work that was popularized in *Scientific American* in 1975. It is interesting to note that Chaitin considers his *ACM Journal* paper of 1975 to be the start of Algorithmic Information Theory and that the early works are just the 'pre-history' of the field (Chaitin, 1999: 88). This is also another good way to 'eliminate' the contributions of both Solomonoff and Kolmogorov to Algorithmic Information Theory. Now while Chaitin is one of the founding three of Algorithmic Information Theory, along with Solomonoff and Kolmogorov, he is the most active of the three in 'popularizing' it to the general public with such works as *The Unknowable* (1999) and *Exploring Randomness* (2001).

As Chaitin has stated in one of his books: 'I publish many, many books and papers on AIT. AIT is my life' (Chaitin, 1999: 86). A.N. Kolmogorov died on October 20, 1987 and only wrote a few early papers on the subject (Young and Minderovic, 1998: 284–286). Solomonoff has written his autobiography (Notes on Artificial Intelligence, Volume 904, Springer-Verlag, 1995, pages 1–18). In this interesting account of his intellectual life, Solomonoff cites Chomsky's 1956 formal language paper titled 'Three models for the description of language' as being ideal for induction (Solomonoff, 1995: 9). Algorithmic probabilies obtained probabilies from compression and Huffman (1952) and information theory codes, Shannon's information theory (1948) that could 'pack' information into bits, use probabilities to compress information (Solomonoff, 1995: 12).

Solomonoff notes the Kolmogorov was interested in the complexity and randomness of strings and the stochastic properties based on length of codes but adds the comment 'Strangely enough, he did not appear to be interested in inductive properties' (Solomonoff, 1995: 18). He mentions Chaitin's work (1966 & 1969) on defining randomness in terms of program length but comments that Chaitin did not investigate the idea of the 'goodness' of a theory in relation to its program (Solomonoff, 1995: 18). Solomonoff cites Schnorr's paper (1973) on 'process complexity' as being the negative logarithm of what he had defined to be Algorithmic Probability (Solomonoff, 1995: 19). Solomonoff uses the term 'brilliant associates' to describe those who had developed algorithmic probability beyond his and Kolmogorov's pioneering work (Solomonoff, 1995: 18).

Solomonoff attended the University of Chicago, 1946–1950, were he later obtained a Master's of Science degree (Li and Vitanyi, 1993/1997: 89). Solomonoff's intent in formulating a general theory of inductive reasoning was to develop a complete system that would overcome those shortcomings found in Carnap's *Logical Foundations of Probability* (1950) (Li and Vitanyi, 1993/1997: 89). Carnap was a lecturer of Solomonoff while he was an undergraduate (Li and Vitanyi, 1993/1997: 89). An interesting note is that Kolmogorov Complexity was an 'auxiliary' concept used to obtain a universal a priori probability that was able to prove the invariance theorem by Solomonoff in his 1964 paper (Li and Vitanyi, 1993/1997: 89–90; Solomonoff, 1964). What comes across in Chaitin's books is his rather 'self-centered' assessment of his place in the history of the founding of Algorithmic Information Theory. Take a look at the following line from his new book *Meta Math* (2005):

> In fact, you can only appreciate Leibniz if you are at his level. You can only realize that Leibniz has anticipated you after you've invented a new field by yourself ... which has happened to many people. In fact, that's what happened to me. I invented and developed my theory of algorithmic information. (Chaitin, 2005: 58)

From a literary stand point, it is clear that Chaitin is using Leibniz as a metaphor for himself. From the fact that Chaitin does not mention either Solomonoff or Kolmogorov as being a co-inventor of Algorithmic Information, it can be judged by the very lack of their inclusion in the book that both Solomonoff and Kolmogorov are to represent 'Newton' to Chaitin's Leibniz.

The fact that Chaitin even calls Newton a 'rotten human being' and that he was 'inferior' as both a mathematician and a philosopher to Leibniz only adds to this speculation (Chaitin, 2005: 57). Chaitin concludes his book, *Meta Math*, with the following passage:

> I think that these observations should serve as an antidote to the excessive egotism, competitiveness and the foolish fights over priority that poison science. No scientific idea has only one name on it; they are the joint production of the best minds in the human race, building on each other's insights over the course of history. (Chaitin, 2005: 142)

I do not know whether to call this passage a comedy or a tragedy' It has the qualities of being both in that Chaitin is both a victim, in that his contemporaries, his peers, have termed it 'Kolmogorov' complexity, and as a perpetrator, in that Chaitin has 'excluded' both Solomonoff and Kolmogorov in the very book he states his 'plea' for a fair and impartial science. A very odd and pathetic testament to the legacy of this 'study of complexity'. Grattan-Guinness's review of Chitin's *Meta Math* book in *Nature* (2006) notes that Chaitin fails to qualify logic from meta-logic and once even 'states it quite wrongly' (Grattan-Guinness, 2006: 791).

Grattan-Guinness also mentions that the book would be served well if some other, non-biological types of complexity, had been used such as 'A.N. Kolmogorov in the 1960s' who had concurrent developments to complexity theory (Grattan-Guiness, 2006: 791). In fact, Chaitin has written *Meta Math* to stand alone with only Chaitin's contributions to the field of Algorithmic Information Theory being included in the development of the ideas behind the theories that gives the impression that it is only Chaitin's theories that have merit. By focusing only on biologically related complexity, Chaitin has excluded both Solomonoff and Kolmogorov's contributions to complexity theory. Grattan-Guiness comments on the style of the book as being better suited to a 'internet chat-room than a book' and concluded the review with the following lines: 'It is nice to have popular books on modern mathematics, logic and science. But it is nicer if they are prepared with care' (Grattan-Guiness, 2006: 791).

Meta Math is also reviewed by Lanier (2006) but Lanier keeps his review focused on the merits of Chaitin's book rather than its idioscracies (Lanier, 2006: 269–271). First, Lanier likes Chaitin's book and, more importantly, he likes his work (Lanier, 2006: 269). Possibly the most telling remark in the book review is the following:

> If there was a prize for books with real live math equations that can hold the attention of reader's who lack technical training, I'd nominate this one. (Lanier, 2006: 269)

Lanier seems to understand that the average person reading a book on 'math' has little, or if American, no experience with even basic mathematics, let alone difficult conceptual models and philosophies being proposed by Chaitin. Perhaps Chaitin is the necessary 'cheerleader' for modern 'mathematical' philosophies that can break into the void called modern living. It is clear that other disciplines, most notably physics, have turned to such stylistic devices as 'pop-culture' allegories, wit, and graphic illustration, sometimes hand-drawn, to present the physical sciences to the lay public (Randall, 2005).

Raatikainen reviews two of Chaitin's books in the *Notice of the AMS* (2001). Raatikainen reviews both *Exploring Randomness* (2000) and *The Unknowable* (1999) both by Chaitin. Raatikainen starts by giving a history to Algorithmic Information Theory by starting with the work of Godel and Turing. He then proceeds to the concepts behind algorithmic complexity. Raantikainen begins his attack on Chaitin's ideas on the philosophical level by noting that Chaitin is 'simply wrong' in that there is no direct dependence between the complexity of an axiom system and its power to prove theorems and that there relevance for the foundations of mathematics has been greatly exaggerated (Raatikainen, 2001: 995).

Raatikainen makes the comment that Chaitin does not respond to criticism of his work but simply evades difficult questions and that his, Chaitin's, writing begins to resemble the 'dogmatism' of his 'opponents' (Raatikainen, 2001: 996). He also calls Chaitin 'megalomaniacal' in some of his pronouncements on Algorithmic Information Theory and that Chaitin is , according to Gacs (1989), trying to present himself as the sole inventor of the main concepts and results of this complexity theory (Raatikainen, 2001: 996). Raatikainen has addressed Chaitin's claims in an earlier paper (2000) and finds Chaitin's results as being 'rather non-dramatic' and are the 'simple consequences of Turing's classical result concerning the undesirability of the halting problem' (Raatikainen, 2000: 218).

Raantikainen has published his doctorial dissertation, titled 'Complexity, Information and Incompleteness' (1998) that re-examines Chaitin's theories behind Algorithmic Information Theory and finds them, on both a mathematical and philosophical level, not all that they are claimed by Chaitin (Raatikainen, 1998). Raatikainen had published a paper, from his work on the dissertation, in the *Journal of Philosophical Logic* (1998) that reiterates what is found in the body of the doctorial work (Raatikainen, 1998a). Chaitin's constant, Omega, is a real number whose digits are equidistributed and which expresses the probability that a random program will halt Omega (Wikipedia, 'Gregory Chaitin'). Raatikainen has raised doubts about the 'genuine' randomness of Chaitin's Omega and that even a 'plausible' definition of randomness that can count such sequences as random and even if the Algorithmic theory of randomness is the most 'perfect' possible theory for randomness (Raatikainen, 2000: 221). One can only imagine what Raatikainen thinks about Chaitin's 'super Omega' (Chaitin, 2006; *New Scientist*, 2001).

Lloyd, in his book on quantum computers (2006), makes the aside: 'Perhaps worse, I inadvertently insulted Gregory Chaitin at a lunch by making a joke about people who believe in the healing power of crystals, unaware that he kept a large crystal in his living room because it helped him concentrate' (Lloyd, 2006: 213). Lloyd makes the comment in his book that Chaitin had originally termed algorithmic information as 'algorithmic complexity' but found the quality of randomness to be the trait found in bit strings with high algorithmic information content (Lloyd, 2006: 188). Lloyd's book is reviewed by Schmidhuber in the American Scientist (Schmidhuber, 2006: 364–365). In this review Schmidhuber notes:

> In fact, Lloyd's belief in true randomness also seems inconsistent with his invocation of Ockham's razor, which favors simple explanations of the universe's history over complex ones. According to both standard and algorithmic information theory, true randomness actually corresponds to maximal information, complexity and description length, the opposite of simplicity. (Schmidhuber, 2006: 364)

Schmidhuber continues:

> The book is least convincing when it comes to the topics of complexity, entropy and algorithmic information. Lloyd compares random events at the quantum level to monkeys typing a random program on the universal computer; this is linked to Ray Solomonoff's basic concept of algorithmic probability theory-namely, that short random programs are more likely than long ones. However, the point of Solomonoff's approach is that some programs can remain short by ceasing to read new input bits. This essential feature seems absent from Lloyd's setup, which demands the permanent creation of new bits corresponding to never-ending programs, thus making each 'program' extremely unlikely. (Schmidhuber, 2006: 364)

Schmidhuber concludes with the statement:

> Generally speaking, the connections between Lloyd's model of quantum processing and algorithmic information theory seem vague. (Schmidhuber, 2006: 364)

This is not to say he is not a founding member of Algorithmic Information Theory's discovery, but rather he has a very 'overwhelming' sense of importance in the 'founding' of this theory, some times at the expense of the other two founding member's contributions. This can also be seen in some of the works of his peers such as John L. Casti's *Five Golden Rules* that states in the bibliography section of Casti's book that Chaitin's book *Information, Randomness and Incompleteness* (1990) tells 'the complete story of Chaitin's independent discovery of algorithmic complexity and its connections with randomness' (Casti, 1996: 222).

Chaitin even thanks Casti in the preface section of his book *Conversations with a Mathematician* (2002) for 'explaining my ideas so well' (Chaitin, 2002: vi). Chaitins book *Information, Randomness and Incompleteness* is a selection of his papers and not new material to the field of complexity. Chaitin, in his book *Exploring Randomness* (2001) states that Solomonoff's early work regarding a theory is more like a program: 'Ray Solomonoff did some thinking along these lines for doing induction', as if Solomonoff's 1960 and 1964 papers did not address these issues up to half a decade before Chaitin (Chaitin, 2001: 18). Perhaps this was done to offset Kolmogorov's legacy as a 'Great Man' in mathematics, one of the last true 'man of all seasons' that contributed to the many fields of mathematics during his life time (Young and Minderovic, 1998: 284–286).

Even Uspensky's (1992) previously mentioned paper failed to cite Chaitin as a founding member of Algorithmic Information Theory. Whether Chiatin's 'self promotion' is a result of this type of revisionist history is similar to asking 'What came first the chicken or the egg?' type of question. There is no easy answer to this type of question. Chaitin has a short biography in his 'collected papers' book (1990) and has a rather slim book that was published as 'Conversations with a Mathematician' that has an 'infectious enthusiasm' for his subject: Algorithmic Information Theory (Chaitin, 2002). There was a 'book review' in *Popular Science* that finds its structure 'bitty' and a 'trifle slim', but still a 'highly recommended book' that 'opens' the way the mathematical minds works (*Popular Science*: Book Review). In the preface to this book (2002) Chaitin states:

> My goal was for this book to be light and bubbly like champagne,
> to show that math and science are fun. Read it and tell me if you
> think I succeeded! (Chaitin, 2002: v)

What more can one say' At least Chaitin gives some credit to the 'other' creators of Algorithmic Information Theory with the line 'in the 1960s I, and independently some other people, came up with some new ideas' (Chaitin, 2002: 28).

My own work in the field of Algorithmic Information Theory/Kolmogorov Complexity is the research subject of this thesis 'The Sub-Maximal Measure of Kolmogorov Complexity'. From the early work on this idea in 1998–2000, an unpublished paper resulted (2001), with the first publication of the idea (2003) from an earlier unpublished manuscript (copyrighted in 2000) I have gradually developed this idea over time to get to this stage of the research. My first introduction to Algorithmic Information Theory came by way of my interest in theoretical linguistics, Chomsky, that in turn, lead to theoretical computer science, and finally to Algorithmic Information Theory (Endnote #6). The *Scientific American* magazine article by Chaitin (1975) was my first true introduction to Algorithmic Information Theory (see Endnote #7). The research for this idea was done at Advanced Human Design that is located in Cupertino, California U.S.A. (see Endnote #17).

Kolmogorov developed complexity theory from information theory. Information Theory was developed over the years 1941 to 1948 by Claude Shannon (April 30, 1916–February 24, 2001) refining his idea for publication in the Bell Laboratories technical journal in a two part article titled 'A Mathematical Theory of Communication' in 1948. Bell Telephone Laboratories are the research and development arm of At&T with a history starting with Western Electric, a subsidiary of AT&T, and was formalized as a laboratory in 1925 (Hugill, 1999: 54). Shannon, with Warren Weaver, would publish the book version in 1949. Warren Weaver was a mathematician and it was he who titled the co-authored work, with Shannon, *The Mathematical Theory of Communication* to emphasis the definitive nature of the work (Solana-Ortega, 2002: 461). Some early works on information theory are Abramson (1963), Ash (1965),

Fano (1961) and Khinchin (1957). Some current publications on information theory are Cover and Thomas (1991), Kahre (2002), and Hankerson (2003). An important collection on information theory papers is Verdu (1999) with the book *Information Theory: 50 Years of Discovery* that contains 25 important papers, mostly from the IEEE Society Information Theory group, that was founded in 1953, and published in their journal. Another collection of information theory papers is the *Claude Elwood Shannon Collected Papers* (1993) published by the IEEE Press and edited by Sloane and Wyner (Endnote #8). It is interesting to note that Shannon, as a boy growing up in Gaylord, Michigan, he worked as a messenger for Western Union (Wikipedia: 1).

Shannon used teletype and the telegraph as examples of discrete channels for transmitting information in his book, with Weaver, *The Mathematical Theory of Communication* (Shannon and Weaver, 1948: 7). It is a sign of the times that Western Union has discontinued their telegram service, as of the end of January 2006, in the face of the growing use of instant messaging and e-mails (*BBC News*, 2006: 1–4). This ends an 'era' that had lasted for over 150 years of service.

The only citations that question Shannon's Information Theory are those found in Gilbert (1966) by Doob (1949, 1959) that raise the question of the mathematical soundness of information theory (Gilbert, 1966: 320; Doob, 1949, 1959). Gilbert's article (1966) is a bit cynical and seems to have all the markings of 'sour grapes' especially in the light that he, Gilbert, is also a member of the mathematical and statistical center at Bell Laboratories in Murry Hill, New Jersey, the same division and company as Shannon, and it takes little effort for Gilbert to find a host of 'problems' inherent in Shannon's information theory (Gilbert, 1966). Joseph Leo Doob (1910–2004) was an American mathematician who specialized in analysis and probability theory (Wikipedia, 'Joseph Leo Doob', 2006: 1–3).

I have yet to see feedback to Doob's two articles, 1949 and 1959, challenging Shannon's mathematical soundness of information theory, especially as both men are well respected in their respective fields. In Doob's 1949 review of Shannon's *A Mathematical Theory of Communication* (1948) in *Mathematical Reviews* (1949) Doob makes the comment 'The discussion is suggestive throughout, rather than mathematical, and it is not always clear that the author's mathematical intentions are honorable' (Doob, 1949: 133). Doob's 1959 Mathematical Reviews citation taken from Gilbert (1966) seems to be a 'phantom' review as none of the 1959 reviews deal with Shannon's work and that both the volume and number to the 1959 edition of *Mathematical Reviews*, Volume 5, Number 3, are for March 1944 (Gilbert, 1966: 326). Gilbert (1966) has raised what seems to be a rather mute point by 1966 in that A.I. Khinchin produced the first mathematically exact presentations of information theory before the 1960s (Reza, 1961: 13; Khinchin, 1957).

Shannon defines 'information' as being a form of entropy (Shannon and Weaver, 1949: 18). The concept and semantics of 'information' is addressed by Solana-Ortega (2002) as it relates to Shannon's original description and presents a case for defining a prune as a dried plum. Solana-Ortega (2002) has sub-titled his paper 'A Homage to Claude E. Shannon' when in fact it is a tribute article to Edward T. Jayynes who revised the definition of 'information' to become the Maximum Entropy Method (Solana-Ortega, 2002: 459). Solana-Ortega (2002) comments that Shannon had originally wanted to define what he would later term 'entropy' as 'information' but thought that it was an over-used term and tried 'uncertainty' but after talking with John von Neumann he settled on the word 'entropy' (Solana-Ortega, 2002: 464). Solana-Ortega (2002) also mentions Bar-Hillel's definition of Shannon's entropy as being the rarity or improbability of the kinds of symbol sequences (Solanan-Ortega, 2002: 464).

Some have criticized the model Shannon used for information theory as it seemed to belong to a 'transmission' model of communication that treated information as a 'commodity' or 'entity' that can be transported from one point to another point along a communications pathway (Solana-Ortega, 2002: 465). Shannon comments in his book (1949) 'The fundamental problem of communication is that of reproducing at one point either exactly or approximately a message selected at another point' (Shannon and Weaver, 1949: 3). Shannon (1949) continues 'The significant aspect is that the actual message is one selected from a set of possible messages' (Shannon and Weaver, 1949: 3). Both Shannon and Weaver (1949) emphasize that 'information' must not be confused with 'meaning', or as Shannon terms it, a 'semantic' aspect to communication (Shannon and Weaver, 1949: 399).

As Singh (1966) reconfirms Shannon's definition as 'the message actually transmitted is a selection from a set of possible messages formed by sequences of symbols of its own repertoire' (Singh, 1966: 12). Singh (1966) continues 'the communications system is designed to transmit each possible selection, not merely the one that happened to be actually chosen at the moment of transmission' (Singh, 1966: 12).

Algorithmic Information Complexity, or AIC, has the following properties (Sommerer and Mignonneau, 2003: 87):

1. The more ordered the string the shorter the program. Less complex.
2. Incompressible strings are indistinguishable from random strings.
3. Most long strings are incompressible.
4. You cannot prove that there are strings above a certain fixed level of complexity using formal systems.
5. In general it is incomputable.

The Algorithmic Information Complexity, AIC, of symbols is the length of the shortest program to produce it as an output (Sommereer and Mignonneau, 2003: 57). Note that the anagram AIC is also used to denote Algorithmic Information Content. Both Sommereer and Mignonneau (2003) state that algorithmic information complexity and computational complexity have meet with great success in measuring complexity (Sommerer and

Mignonneau, 2003: 57). Computational complexity tries to classify solvable problems according to their intrinsic computational difficulty (Sommerer and Mignonneau, 2003: 88).

It was not until that late 1960s and early 1970s that 'structural complexity theory' developed by Cook (1971) and Karp (1982) allowed for this type of development (Sommerer and Mignonneau, 2003: 88–89). Selman (1986) edits the proceedings from the 'structure in complexity theory' that took place at the University of California in 1986. The first person to address how difficult it is to compute some function was Rabin (1959, 1960) (Jones, 1997: 24). Blum (1967) would introduce a general theory of complexity independent of any specific model of computation (Jones, 1997: 24).

Concepts and terminology on transmission errors in information theory can be found in Hankerson, Harris, and Johnson (1998) and Klir (2006). Information and coding theory can be found in Jones and Jones (2000). Early work on transmission theory can be seen in Nyquist (1924) and Hartley (1928). Hartley is given credit for being the first to use the term 'information' in the technical sense in his 1928 paper (Millman, 1984: 48; Hartley 1928).

Hartley studied the basic relationship between the width of frequency band and the capacity of a system to transmit that 'information' (Fagen, 1975: 909). Nyguist's work with the behavior of digital and analog signals resulted in major contributions to the advancement of transmission technology (Fagen, 1975: 766). Foundational work on the properties of 'noise' in an electrical system was done by J.B. Johnson and published in *Physical Review*, Volume 32, Number 97 in 1928 (Williams, 1978: 1248).

But it would be Claude Shannon, hired in 1941 by Bell Laboratories, for his ground breaking Master's thesis that used use Boolean algebra and symbolic logic for the synthesis, analysis, and optimization of relay circuits, to take information theory to the summit of development (Fagen, 1978: 165) and (Endnote #9). Reza (1961) notes that A.N. Kolomogorov has contributed to the development of information theory and that A.I. Khinchin produced one of the first mathematically exact presentations of information theory (Reza, 1961: 13). Reza (1961) remarks that work of E.C. Cherry had attributed the early development of telecommunications as being a part of one of the fields that helped shape the development of information theory (Reza, 1961: 11). Needham (1996) makes the following comment on the relationship between the computing industry and the communications industry:

> The communication industry is used to regulating heavy long-term investment: high utilization of plant, striving for good uniform

quality of service. The computer industry is the most unregulated there is; it sees five years hence as infinity; most of the things it sells are heavily underutilized; it depends on maximum speed to market. (Needham, 1996: 291)

The following result is noted by Needham (1996):

To the computer person the communications industry is ponderous, slow, overcautious; congenitally fussy. To the phone company man the computer industry is archaic, has a throw-away culture, is wasteful, peddles unreliable trash. (Needham, 1996: 291)

Shannon makes the comment in his book *The Mathematical Theory of Communication*, in a footnote, that communication theory was 'indebted' to Norbert Wiener for much of its basic philosophy and theory (Shannon and Weaver, 1949: 52) and (Endnote #10). Shannon cites Wieners (1949) early work on communication theory as being a statistical problem and Wieners' work (1948) on communication and control in general (Shannon and Weaver, 1949: 52–53). Wiener would later be known for the concept of 'cybernetics' and would write a treaty on concerns over technology and human freedom (1950). Some comments on Wiener's cybernetics and social organizations can be found in Masani (1989: 330–331).

Wiener wrote in the definition to the entry 'Cybernetics' in the Encyclopedia Americans as 'a word coined by Norbert Wiener' that according to Watanabe (1985) had been actually used by Andre Ampere a century before Wiener and used by Plato even earlier in describing processes that 'influence' and 'control' a natural order (Wiener, 1985: 215 and 804). Wiener cites his work in communications engineering as being similar to Shannon's work (1948) as far as results are concerned (Wiener, 1949/1985: 198). Mindell (2002) contends that the convergence of communication and control predated Wiener's cybernetics in the form of interwar engineering developments: gunfire control, aircraft and ship control, communications engineering, and early control theory (Mindell, 2002: 120). Wiener articulated cybernetic theory, but did not originate it (Mindell. 2002: 120). Wiener failed to cite in his works the early developments of Elmer Sperry, Nicholas Minorsky, Harold Black, Harry Nyquist, Hendrick Bode, or Harold Hazen, all of which predate Wiener's cybernetics (Mindell, 2002: 286).

Chaitin would use a similar technique to 'remove' the competition of a 'original' idea many decades later. Wiener cites Kolmogoroff's 1941 paper as paralleling his own work that started at about the same time and was done for the National Defense Research Committee, section D, while at MIT (Kolmogoroff, 1941; Wiener, 1949: 59; Wiener, 1958: 126). Wiener would later write about his childhood, he was a gifted child, in his autobiography (1953) and his adult years (1956). Conway and Siegelman (2004) have written a biography of Wiener that covers this unusual man's life.

Cybernetics would by the 1960s fall out of favor, at least in the United States, and be replaced by other scientific 'fads' such as 'catastrophe theory' (Thom, 1975) that would, ultimately, be replaced by newer theories (Horgan, 1996: 207–208). Does anyone remember 'fuzzy logic?' (see Endnote #11). Even 'chaos theory', given wings by the book by Gleick (1987), was given a re-evaluation by one of its founders, Ruelle (1991), with hopes that it will again find some legitimacy in academic circles (Horgan, 1996: 208–209).

The current scientific battle is with Stephen Wolfram's book *A New Kind of Science* (2002), that has Wolfram, an 'overall boy wonder', accused of 'not being meticulous' with crediting other peoples contributions to Wolfram's work (Wright, 1988: 62). In this case it is Ed Fredkin, who's work predates Wolfram's by decades, but who has published little in the area of 'cellular automata' that was developed from the early work of von Neumann and Stanislaus Ulam (Fredkin and Toffoli, 1982). Some reviews on Wolfram's *A New Kind of Science* commented on the size of the book, 1197 pages in length, (Matthews, 2002), while others called his work a 'silly book' (Derbyshire, 2002: 1) because of Wolfram's vanity, carelessness, and error in the production of his 'self-published' book (Derbyshire, 2002: 4).

Schmidhuber (2003) in a letter to a physics journal has claimed that Wolfram has 'borrowed' earlier ideas of the concept of the universe as a computer program from Konrad Zuse, of early computer fame, who published his ideas in a 1967 paper (Schmidhuber, 2003: 2). Aaronson (2001) in his review of Wolfram's book finds that the end notes, of which there are 349 pages worth, make better reading than the main text of *A New Kind of Science* and that when Wolfram cites original material to his topic, he is quick to call it 'misguided' or 'irrelevant' to his scientific 'discovery' (Aaronson, 2001: 96).

Weinberg (2002) perhaps raises the most important question about Wolfram's work: 'why hasn't the problem of defining complexity been stated, let alone been proved' (Weinberg, 2002: 9). This is a fundamental question to ask about any work, let alone a 'new science'. Gray (2003) reviews Wolfram's work and states that Wolfram's focus is on 'discrete' systems, like cellular automata, and while being impressed with the scope of the book, is not convinced that it is a new science (Gray, 2003: 200–201). Many of the reviewers that had reviewed the book used the anagram 'ANKS', for *A New Kind of Science*. I guess to save on ink and paper.

Some serial publications on Algorithmic Information Theory/Kolmogorov Complexity are as follows: Algorithmic Learning Theory (Lecture Notes in Computer Science), a bound proceedings of the Algorithmic Learning Group, usually an annual event, The International Series in Engineering and Computer Science (Springer-Verlag), Texts in Theoretical Computer Science, An EATCS Series, also by Springer-Verlag, Cambridge Nonlinear Science Series (Cambridge University Press), Studies in the Sciences of Complexity (Addison-Wesley Publishers), and Texts in Computer Science (Springer-Verlag). Dissertations on Algorithmic Information Theory/Kolmogorov Complexity are a growing category: Ronneburger (2005) from a dissertation titled 'Kolmogorov Complexity and Derandomization' that was awarded in 2004 by Rutgers, the State University of New Jersey and examines different notions of resource-bounded Kolmogorov complexity. Popel's (2000) dissertation is on 'Information Theoretic Approach to Logic Functions Minimization' at the Technical University of Szczecin, Szczecin, Poland. Sow's (2000) Ph.D. dissertation from Columbia University is titled 'Algorithmic representation of visual information'.

Note: I was unable to review the book *Kolmogorov Complexity and Randomness* (North-Holland Mathematical Library Series) because it will not be published until November 1, 2006. Also *Algorithmic Information Theory: Mathematics of Digital Information Processing* by Peter Seibt (Springer-Verlag) will not be published until November 2006.

Huffman Coding, Huffman's algorithm, was discovered in 1951 by David Huffman, a graduate student at MIT, published in 1952, while he was taking a class as a student of R.M. Fano (Hankerson, Harris, and Johnson, 1998: 107). History has it that Fano gave the problem to the class without telling them that the problem had no solution (Hankerson, Harris, and Johnson, 1998: 107). Solomonoff mentions that Huffman obtained a short code from the knowledge of probabilities and he, Solomonoff, obtained probabilities from the knowledge of short codes (Solomonoff, 1995: 6). Shannon's 1948 paper introduced algebraic coding theory, devised by R.W. Hamming, as well as his developments from the work of R.E. Hersey in the form of Hersey's '2-out-of-5' code of 1938 (Millman, 1984: 52).

Brillion (1956/1962) notes that a Hamming code is used to detect the position of an error and to correct it (Brillion, 1956/1962: 63; Hamming, 1950). Shannon was apparently unaware of these early 'cable codes' that had the principles of 'error correction', but did not apply them, even thought error detecting codes had been used in cable telegraphy since the 19th century (Millman, 1984: 52; Friedman, 1928; Friedman and Mendelsohn, 1932). Many different systems had been suggested by the late 1930s and during the war years. M.E. Mohr had suggested a 'tertiary' or three level system as being the most efficient for coding (Fagen, 1978: 316).

During the war H.L. Barney carried out research on various possible combinations, 2, 4 and 8 valued signals, for Bell Labs and the U.S. Government, but it would be a competitor, International Telephone & Telegraph Company, ITT, that had already used, and patented; French patent 1938 and U.S patent 1942, a binary coding system (Fagen, 1978: 316; Barney, 1945; Reeves, 1965). But the inventor for ITT, Alex H. Reeves, was in the war effort, and the research was handed over to Bell Laboratories in 1943 (Fagen, 1978: 316). Takahashi (2004) shows a asymptotically universal code that is less or equal than that of the minimal description length, MDL, code (Takahashi, 2004).

Some current studies in compression are Davisson and Gray (1975, 1976), Loreto and Puglist (2003), Oexle (1995), Wolfowitz (1978), Purser (1995), Gagie (2006) with Zayed (1993) dealing with Shannon's sampling theory. Compression is the process of reducing the number of information bearing units, usually bits, from the original information source without the lose of the original information content. Data compression is done with the use of codes; encoding to compress, decoding to decompress.

Because most data used in actual applications have a statistically high level of redundancy a lossless compression factor can be achieved using such compression algorithms. Lossy compression affords a lose of data but not at the expense of the complete contextual content of that data to be compressed. Lossless compression can be reconstructed completely from compress where as lossy compression cannot be resurrected from a compression state.

The development of communication techniques for algorithms has a history before its rapid growth in the twentieth century. Schreiber (2003) states that Jordanus de Nemore, circa 1200 A.D., was the first person to use letters as variables for given and required quantities, but it was the work of Francois Viete in the 1500s that had lasting influence (Schreiber, 2003: 688). In the 1600s a bountiful algorithm oriented computational tradition developed with Robert Recode of England, Adam Ries and Ulrich Wagner in Germany, and Niccolo Tartaglia in Italy (Schreiber, 2003: 689). A development from the 'Ars Magna' of the 13th century logican the Spaniard Ramon Llull was the use of artifical languages of a universal expressive power and machines and systems of rules to decide truth or propositions and solving problems that would later influence Descartes and Leibniz (Schreiber, 2003: 689).

The history of computing can be found in Flamm (1988). The development of Silicon Valley is addressed in Lecuyer (1999). The history of computing is long and varied but some individuals stand out in the development of core ideas to the foundations of computing. Gottfried Wilhelm Leibniz (1646–1716) was the first to formulate the basis of modern symbolic thought through his studies of binary arithmetic as well as his work on early calculating machines (Ifrah, 2001: 251; Williams, 1997: 129–136). George Boole (1815–1864) used logic to give certain premises, or conditions, that determine the predicates of a class of objects subject to those conditions (Ifrah, 2001: 253). Boole also to join the procedures of propositional logic within the operations of a true algebra (Ifrah, 2001: 251).

Boole published his ideas in two books, *Mathematical Analysis of Logic* (1847) and *The Laws of Thought* (1854). Symbolic logic is the science of human thought establishes a system of axioms as well as the rules and procedures which govern the various relations between the repositions and check the consistency, the compatibility, and the independence of the axioms that have been postulated (Ifrah, 2001: 269). Although Charles Dodgson (1832–1898), a.k.a. Lewis Carroll, was a contemporary of Boole, and although he is known for being a logician, Dodgson was not a major contributor to the field of logic (Gillispie, 1971: 138). It is not even clear that Dodgson had read Boole's *Laws of Thought* (1854) even though Dodgson owned a copy of the book (Gillispie, 1971: 138). It would be Claude E. Shannon who, in 1937, wrote his Master's thesis on 'A Symbolic Analysis of Relays and Switching Circuits' and proved that the rules of Boolean algebra could be applied to electric circuits and that these circuits could perform the fundamental operations of the algebra (Ifrah, 2001: 257; Shannon, 1938).

The use of Kurt Godel's findings that the 'undecidability' of certain general axiomatic theories, in a 1931 paper, lead Alan Turing to confirm Godel's findings in that reducing mathematical reasoning to symbolic calculation would lead to conclusion that it would be impossible to find a logical sequence of elementary operations sufficiently general to determine whether a given theorem is demonstrable (Ifrah, 2001: 279–280; Nagel and Newman, 2002).

Turing's mathematical notion of a 'universal algorithmic automation' would be the theoretical model for all computers of the future (Ifrah, 2001: 292). The 'synthesis' of the modern computer would take root in the development of the construction of the analytic calculator: the EDVAC, Electronic

Discrete Variable Automatic Computer, in 1944 with the famous mathematician John von Neumann (Ifrah, 2001: 281). According to Ifrah, the single most important factor in the development of the modern computer, to this point in time, is the development of the 'stored programme' (Ifrah, 2001: 281).

This was described in von Neumannn's foundational paper of June 30, 1945 titled 'First draft of a report on EDVAC' (Ifrah, 2001: 281). Goldstine (1972) states that this 'First Draft' paper (1945) by von Neumann was a working paper for clarifying and coordinating the thinking of the project group and not intended for publication (Goldstine, 1972: 196). A look back at what computing was and what it would be in the 'future' can be seen in the 'dated', but interesting, work of Berkeley (1949) and Dreyfus (1972). The History of Computing series by The MIT Press, edited by I. Bernard Cohen and William Aspray, has a list of titles dealing with the growth and development of early computing (see Endnote #12). Agar (2006) reviews Copeland's book (2006) on the secret code breaking computers of 'Bletchley Park' during World War Two (Agar, 2006: 746; Copeland, 2006).

Current works on computers and computability are Floyd and Beigel (1994), Lassaigne and de Rougemont (2004), Hromkovic (2004) and Jones (1997). Johnson (2003) describes the building of a chess playing computer in the mid-1970s by a group of computer science students using more than a hundred 'Giant Engineer Tinkertoy' sets (Johnson, 2003: 20). My own creative experiences began with tinker toys, play doh and legos as a child and it seems that this 'creative drive' is still with me into my forties (Los Altos News, 1965; du Sautoy, 2006).

The development, or rather research into, quantum computing is addressed in Milburn (1998), Meyer, in Lomonaco (2002), Brooks (2003), Hey and Allen (1999), Deutsch (1997), Brown (2001), Terhal, Wolf, and Doherty (2003), Wolfram (2002), Pavicic (2003), Dowling (2006), Lloyd (2006) and Cho (2006). Quantum Kolmogorov Complexity is addressed in a paper by Berthiaume, Van Dam, and Laplante (2000) and by Vitanyi (2001). Quantum Algorithmic Entropy is the topic of Gacs' paper (2005). Woesler (2005) attempts to solve the Copenhagen interpretation of quantum theory with Kolmogorov complexity.

Le Bellac (2006) has published *A Short Introduction to Quantum Information and Quantum Computation* (Cambridge University Press). Information distance is the subject of a paper by Bennett, Gacs, Li, Vitanyi, and Zurek (1993). An 'information-theoretic' approach to neural computing is addressed in Deco and Obradovic (1996). An article in the *New Scientist* questions whether quantum computers can ever over come noise in the system (*New Scientist*, 2006: 17). Svozil (1996) addresses quantum algorithmic information theory (Svozil, 1996). Wheeler (1994: 295) addresses a new view of reality in 'It from Bit' from his autobiography.

Chaitin uses McCarthy's LISP program in his proofing methods of Algorithmic Information Theory, most notably in his book *Algorithmic Information Theory* (Cambridge: Cambridge University Press, 1987). Chaitin discovered the LISP programming language in 1970 while he was living in Buenos Aires and has made it his programming language of choice with *Algorithmic Information Theory* (Chaitin, 2005: 46). Six of Chaitin's books use LISP and he notes that the structure of LISP, using a few powerful, but simple, basic concept to make it a 'practical tool' for his applications (Chaitin, 2005: 45–46). LISP was developed by McCarthy in the Summer of 1956 through the Summer of 1958 and during the time when it was implemented in addressing problems of artificial intelligence, from the Fall of 1958 to 1962 (Wexelblat, 1981: 173) and (ACM Press, 1987: 258).

LISP was developed by universities in the late 1950s and early 1960s by the U.S. government, through DARPA, as financial support for the study of artificial intelligence (Flamm, 1988: 26; see also Endnote #13). LISP uses a recursive manner of conditional expressions with the representation of symbolic information external, in the form of lists, and internally by list data structures (Ralston, Reilly, and Hemmendinger, 2000: 991). Computer programming is both an art and a science (see Knuth), but can be defined by the school of programming known as 'structured' programming that was developed by Konrad Zuse, best known for designing and building 'computers' for the German's during World War Two (Flamm, 1988: 159; Gutknecht, 1990: 305). Wirth (1976), Dijkstra and Hoare (Dahl, Dijstra, and Hoare, 1972) are renowned computer scientists from the 'structured' programming school of thought (Wirth, 1976: xii).

Structured programming is defined as a methodological style where a computer program is constructed by concatenating or coherently nesting logical subunits that either are themselves structured programs or are of the form of one of a small number of clear control structures (Ralston, 2000: 1701). It is interesting to note that Dijkstra (1982) makes the comment that in the LISP 1.5 Manual the authors give up half way through the description of their programming language and then try to compliment their incomplete language definitions by an equally incomplete sketch of a specific implementation (Dijkstra, 1982: 64; McCarthy, 1962). For an insider's account of the early year's of computer programming see John Backus's personal account of the development of the programming language FORTRAN in Metropolis, Howlett, and Rota's A History of Computing in the Twentieth Century (1980). Backus mentions in the article that 1950s America was 'untainted' by scholarship or academia and had a 'vital frontier enthusiasm' for the field (Metrolpolis, Howlett and Rota, 1980: 126).

The early development of computers had the question of whether an 'analogy', or measurement, machine or a 'digital' computer would be the computer of the future (Goldstine and von Neumann, 1946: 4). In that paper von Neumann comments that the digital machines up to that point have been decimal although some new binary machines would be built and that analogy or 'measurement' computers being left out of the discussion (Goldstine and von Neumann, 1946: 8). In the second paper von Neumann comments that although digital machines have been using the decimal system, von Neumann felt that a binary system should be used for the next generation of computers (Burks, Goldstine, and von Neumann, 1946: 41).

Another important book on early computing, this includes early artificial intelligence work, is Shannon and McCarthy's *Automata Studies* (1956) that has important papers from de Leeuw, Shannon, McCarthy, and other important researchers from this era (Shannon and McCarthy, 1956). Baldwin and Clark (2000) find modularity in the early design of computers in that modularity does not arise by chance but rather the intentional outcome of a conscious design effort (Baldwin and Clark, 2000: 249). This is a 'pre-ordained' quality from the designer's of such systems and is the result of a 'steady accreditation' over many design cycles (Baldwin and Clark, 2000: 249). Baldwin and Clark cite the development of the first modular computer system the IBM System/360 of the 1960s (Baldwin and Clark, 2000: 169–194).

Knuth's *The Art of Computer Programming: Volume 2 Semi-Numerical Algorithms* (1998) was the primary source for the description and history of binary, ternary and quaternary systems (Knuth, 1998: 194–328). Knuth cited Leibniz's *Memoires de l'Academie Royale des Sciences* (Paris, 1703: 110–116) as bringing binary notation into the 'limelight' (Knuth, 1998: 200). The history of binary notation systems is detailed in Anton Glaser's *History of Binary and Other Non-Decimal Arithmetic* (Los Angeles: Tomash, 1981). Knuth cites George R. Stibitz as developing an excess 3 binary coded decimal notation in the early development, the 1930s, of 'general arithmetic operations' for electromechanical and electronic circuitry (Knuth, 1998: 202).

Knuth cites Brian Randell's book *The Origins of Digital Computers* (Berlin: Springer, 1973) as a excellent source of reprints of early papers (Knuth,

1998: 202). Knuth draws attention to W. Buchholz's paper 'Fingers or Fists' (*CACM*, Vol. 2, [December 1959]: 3–11) as a retrospect work on the merits of binary systems in computing from von Neumann's detailed suggestion in the 1940s (Knuth, 1998: 202). Buchholz notes that the number system used is a major factor in the arithmetic speed of a computer (Buchholz, 1959: 3). The earliest known digital electronic circuit was described by Eccles and Jordan in 1919 but it was W. Bryce of IBM that investigated early applications to electronic 'calculating machines' in 1915 (Randell, 1982: 293). Wynn-Williams used thyratrons in binary counting circuits at the Cavendish Laboratory at Cambridge in 1932 (Randell, 1982: 293).

Randell (1982) has included E. William Phillip's 1936 paper 'Binary Calculation' that was a call for a binary system in the actuarial profession (Randell, 1982: 303–314). Phillip's makes the clarifying point that Leibnitz did not, in fact, 'invent' the binary arithmetic as it was attributed to a Chinaman; Fohi, 23rd century B.C, who developed the Cora or binary system (Randell, 1982: 306). Glaser (1971) makes the comment that William Ernst Tentzeln, editor of *Curieuse Bibliotheca* (1705), considered it odd that the supposedly intelligent Chinese had lost and then failed to 'rediscover' the meaning of the Figures of Fohy, and that it took a European genius, Leibniz, to do the job for them (Glaser, 1971: 49). Glaser (1971) also makes the comment that up to 1900 very little had been mentioned about binary and other nondecimal numeration (Glaser, 1971: 115).

The famous 20th century mathematician Richard Courant noted in his book 'What is Mathematics' (1941) that a base 4 works 'best' because it requires the least number of concepts and names (Glaser, 1971: 125). Number representations in early computing included excess-three and boundary codes that are 10, 4, 4, and 7 bits long (Glaser, 1971: 139). Glaser notes that a 3-bit code is not possible in that there is only eight 3-bit strings from 000 to 111 (Glaser, 1971: 139). Glaser (1971) notes that decimal codes greater than four usually involve a 'parity code' that involves no additional information and is given to add an odd parity to the string (Glaser, 1971: 146).

This is also known as a 'redundancy' bit (Glaser, 1971: 146). Gilbert (1966) notes that a parity check is a constraint requiring that the sum of the digits in certain positions be an even number (Gilbert, 1966: 324). Shannon (1950) mentions that computers of numeration systems could use negative digits, especially if the radix was odd so as to be symmetric, with the negative digits equaling the number of positive digits (Glaser, 1971: 160). Richards (1955) notes that that radix three, a ternary system, os the most efficient and that the radix two and radix four are less efficient that radix three (Richards,

1955: 8–9).

Richards (1955) also notes that the radix three has less advantages that radix two when balanced with then current, 1955, computer components (Richards, 1955: 25). Richards (1955) makes the comment that no new ideas have been employed for the zero, one and two for ternary systems (Richards, 1955: 3–4). Richards has noted the types of notation used for ternary systems (Richards, 1955: 4):

1. A series of three 1's.
2. A new symbol such as 3.
3. A 1 in the prefix position followed by a 1 such that they are interpreted as two plus one.

Richards (1955) also mentions negative numbers (Richards, 1955: 14–15):

1. A -1, 0, and $+1$ are used instead of 0, 1, and 2.
2. These symbols, -1, 0, and 1, can be abbreviated to $-$, 0 and $+$.

Brillion (1956/1962) mentions the ternary system in the coding section of his book and lists a 'possible ternary code for the letters of the English language' (Brillion, 1956/1962: 53). Brillion notes that the ternary system used has positive and negative aspects and uses the ternary code: -1, 0, $+1$, that is presented in Table 5.3 (Brillion, 1956/1962: 53–54). Brillion (1956/1962) remarks that the ternary code is not balanced as the symbol -1 is less frequent than 0 and 1 (Brillion, 1956/1962: 54).

Richards (1955) concludes that with the dearth of ternary computer components and the design problems of such a ternary computer system, the disadvantages of such a computer far out weighs the advantages (Richards, 1955: 14–15 and 25). A 'balanced ternary' system is a non-standard positional numeral system that is ideal for comparison logic and is used in computing (Wikipedia, 'Balanced Ternary' 2006: 1). Early Russian experimental computers used a balanced ternary system (Wikipedia, 'Balanced Ternary' 2006: 2). Weinstein (2003) notes that Knuth (1981) finds no substantial application of a balanced ternary notation has been made to computing (Weinstein, 2003: 2961). A common usage for a ternary system is in American baseball that uses it to denote the fractional parts of an inning in a baseball game (Wikipedia, 'Ternary Numeral System' 2006: 1).

In *Arithmetic Operations in Digital Computers* (1955) Richards states that using a binary system for printing produces problems because of the numerous amounts of 1's and 0s that result in excessive errors (Richards, 1955: 5). Ding, Kohel and Ling (2000: 285) mention in their paper the work of Ding, Kohel, and Ling (2000) that uses a class of ternary codes for a secret-sharing scheme. Zhao and Sham (2001: 1) use a binary and mixed radix based algorithms to gene counting procedures. In Zhao and Sham (1998: 225) the authors use a 'ternary' based system for calculating probabilities in determining twin zygosity. Gillie (1965) reviews binary and ternary based systems.

The literature on 'information' is vast and usually written in a 'popular' or 'general public' type of writing style. Seminal works on 'information' are Szilard (1929), Nyquist (1924), and Hartley (1928). Brillion (1956/1962) notes that Szilard's paper of 1929 was the first to tie together the notion of information and entropy (Brillion, 1956/1962: xi and 176). Lanouette (1994) notes that Szilard's paper of 1929 was written in 1922 as a second paper from his doctoral thesis that extended his work on thermodynamic equilibrium from a physical phenomena to the types of activities these would perform as 'information' (Lanouette, 1994: 63).

Lanouette (1994) comments that information theories early developers probably did not know of Szilard's 1929 paper and it was most likely the intervention of John von Neumann, who knew both Szilard and Shannon, to urge Shannon to use the term 'entropy' for his concept of information (Lanouette, 1994: 64 and 332). Brillion's *Science and Information Theory* (1956/1962) concludes that information and physical entropy are of the same nature in that entropy is the measure of the lack of detailed information about a physical system (Brillion, 1956/1962: 293). Brillion (1956/1962) notes that a feedback system uses only positive feedback from a physical system so information in such a system is just the 'value' of that information (Brillion, 1956/1962: 296). A current book by Seife (2006) has the the ungainly title of *Decoding the Universe: How the New Science of Information Is Explaining Everything in the Cosmos, from Our Brains to Black Holes.*

A review of the book in *The New Scientist* also has noted the 'hyped' title but finds the book 'an excellent job' in writing on a difficult subject (Buchanan, 2006: 47). Another current publication on information is by von Baeyer (2004). I have used both Horgan (1996) and Wright (1988) to give a general overview of the state of information sciences in the late 20th century. Probably one of the earliest book I read on information was Campbell's book (1982) that I read over twenty years ago while I was an undergraduate at college and found the book very interesting at the time. MacKay (1969) is a review of early papers on information, while Simon (1969) examines why complex systems have subsystems developed from even smaller subsystems (see Endnote #14). Malescio (2006) reviews Coles (2006) book on probability theory and its role in science (Malesco, 2006: 918; Coles, 2006).

Landauer (1961) was the first paper to study the thermodynamic cost of erasing information and represents real reversible computing devices (Li and Vitanyi, 1997: 586–587; Landauer, 1961). Zurek (1991) edits papers on the relations between physical entropy and Kolmogorov complexity (Zurek, 1991). Solana-Ortega (2002) is an overview of Shannon's work on the 'information revolution' (Solana-Ortega, 2002). Machta (1999) addresses entropy, information and computation (Machta, 1999). An ever expanding area of the use of Shannon's information theory is in the vague field of 'management science' that uses legitimate science in a dubious manner to fit business models. A fine example of this art is Luenberger's (2006) 'Information Science' (Luenberger, 2006).

Some further developments from Solomonoff, Kolmogorov, and Chaitin in Algorithmic Information Theory/Kolmogorov Complexity have been Levin (1974) in discovering, simultaneously with Chaitin (1975), redefining the main theorem and Solovay (1975) and Gac (1974) in developing beyond some boundary definitions by Chaitin (1975). Mutual, or common, information in Algorithmic Information Theory has been emphasized by Fine (1973). Work using DNA sequencing compression using Algorithmic Information Theory can be found in Milosavljevic and Jurka (1993), Chen, Kwong and Li (1999) and Powell, Dowe, Allison and Dix (1998). Hartmanis (1983) has considered the amount of work, time complexity, involved in reconstructing the original data from its description (Barzdin, 1988: 142).

In using an on-line data base search, Expanded Academic ASAP Plus, using Algorithmic Information Theory and Kolmogorov Complexity as the two fields, the following list of current citations resulted: Kreinovich and Longpre (1998), Lateva, McGill, and Pajuman (1998), Shen (1999), Chen and Yeh (2000), Romaschchenko, Shen, and Vereshchagin (2002), Shen and Vereshchagin (2002), Muchnik (2002), Wang (2002), Grunwald and Vitanyi (2003).

Grunwald and Vitanyi (2003), Kurtz (2002), Wang (2002), Gacs, Tromp and Vitanyi (2001), Vitanyi (2001), Gacs (2001), Soklakov and Schack (2000), Raatikainen (2000), Machta (1999), Calude and Chaitin (1999), Shen (1999), and Raatikainen (1998). Vyugin (1998 and 1998a), Garbanzo (1998), Sorensen (1998), Hoffmann (1997), Enamullah, Renz, El-Ayaan, Wiesinger, Linert, and Hoffmann (1997), and Ford (1989 and 1989a). It is interesting to note that when using the ScienceDirect database for a search resulted in no entries when Algorithmic Information Theory was used but had 191 articles when Kolmogorov Complexity was used for the search. Lambalgen

(1989), Kalnishkan, Vovk and Vyugin (2005) and Malyutov (2005) are papers discussing aspects of algorithmic information theory.

Secondary works that discuss Algorithmic Information Theory/Kolmogorov Complexity in a general manner are Beltrami (1999), Bennett (1998), Berlinski (2000), Campbell (1982), Casti (1996), Franzen (2005), Gell-Mann (1994), Goldstein (2005), Seife (2006), Shankar and van Rijsbergem (1997), Siegfried (2000) and Lloyd (2006). Casti and Karlqvist (2003) edit Art and Complexity that seems to involve aspects of neither (Casti and Karlqvist, 2003).

Chaitin has a long list of references that can be divided up into the following groups: the early years 1966–1975, the journal years, 1975–1985, the popular press years 1985–1995, and current publications 1995–2006.

In reviewing the growth of Algorithmic Information Theory/Kolmogorov Complexity as a distinct discipline, I am wondering if it will, or is currently, suffering from, what Claude Shannon termed 'The Bandwagon', by an article of the same name (1956), about the 'improper' interdisciplinary use, or misuse, of information theory in less than a decade after it was invented by Shannon himself? Most theories seem to go through a 'fashionable' phase and although Algorithmic Information Theory/Kolmogorov Complexity is in its fourth decade of existence it seems to be going through a 'late' development stage as a 'popular' method of analysis. Murkowski (1997) states that Algorithmic Information Theory may have implications to legal and economic systems, taking that same pathway as did information theory in the 1950s into areas that have little or no reason for such justifications (Markowski, 1997: 22).

In Grunwald, Myung, and Pitt's work on Minimum Description Length, or MDL, that was a development from Algorithmic Information Theory/Kolmogorov Complexity, the focus of the book is on developments on Minimal Description Length started by Jorma Rissanen in a paper from 1978 (Grunwald, Myung, and Pitt, 2005: 17). The only citation to the founding three of algorithmic complexity theory is on page 17 of this text (Grunwald, Myung, and Pitt, 2005: 17).

In a paper by Small and Tse (2002) using MDL, Minimum Description Length, in neural networks for time series prediction, only cites Rissanen as a historical footnote, with no belabored account of the 'legacies' of MDL's foundations (Small and Tse, 2002: 066701-1). In some respects, Algorithmic Information Theory/Kolmogorov Complexity is moving beyond its foundational nature and it would not be surprising to see no references to Solomonoff, Kolmogorov, or Chaitin in future publications on aspects of algorithmic complexity theory.

Inspiration for Rissanen's work (1978) on Minimal Description Length came from Kolmogorov's 1965 paper and Akaike's 1973 seminal paper on algorithmic information criterion method for model selection (Grunwald, Myung, and Pitt, 2005: 17). Minimal Description Length, MDL, is related to Minimal Message Length, MML, developed by Wallace, along with other authors, without the knowledge of Kolmogorov complexity (Grunwald, Myung, and Pitt, 2005: 17–18; Wallace and Boulton, 1968, 1975; Wallace and Freeman, 1987). Grunwald, Myung, and Piit's book cites Solomonoff's 1978 paper that extends the early work done on Minimal Description Length, MDL, to form an 'idealized' version of MDL (Grunwald, Myung, and Pitt, 2005: 7).

The stochastic processes, such as random and non-random properties, found in computer science are best addressed in Knuth's *The Art of Computer Science* with emphases on Volume Two 'semi-numerical Algorithms' (Knuth, 1981). Knuth (1981) makes the comment that it is not easy to invent a fool proof source of random numbers and that even the definition of 'randomness' is avoided by mathematics and statistical fields by stating the 'how' rather than the 'what' of such processes (Knuth, 1981: 4 and 149).

Works on probability theory by Kolomogorov (1933/1956) and Doob (1953) are also suggested by many authors. In this thesis randomness of a word will be defined as any description of its creation is at least as large as its full representation bit by bit (Hromkovic, 2004: 44). Non-randomness of a word is any description of its creation is less large than at its full representation bit by bit.

On the subject of numbers as words and symbols the following has been researched. Schillinger (1976: 38) examines the connection between mathematics and art and finds uniformity being a primary foundation of this unity and notes the concept of the natural integer as being fundamental. Menninger notes that the number 3 is the first step towards infinity and that it has the relation to the concept of 'the many' away from the notions of 1 as 'I' and 2 as 'you' in an anthropological sense of cultural numerology (Menninger, 1969: 16–17). In English the word closes to number expressions in its distributional properties is the English word 'many' (Hurford, 1975: 3).

Menninger (1969: 147) makes an interesting note that the initial constant [f] in the Gothic word fidwor, 'four', was not originally in the phonetic shift but rather a gradual erosion by constant repetition due to the fact that the number four (fidwor) is followed by five (fimf) and in time adapted to the rhythm of the initial consonant for five (fimf): [f]. Menninger suggests that the number four is a linguistic derivation for 'tip' from the four sides of the cross (Menninger, 1969: 148). That words would have numeric origins can be seen in such words as 'tribute' that was derived from the Latin 'tri-bus' that meant 'third' or 'third part' and then a 'district' then a 'community' that inferred a group or tribe of commonly related peoples (Menninger, 1969: 177).

The word four (4) has always been used to note a 'square' with the basic form being 'quadratus' that Albrecht Durer used in coining the German word 'Vierung' (Menninger, 1969: 178). Butterworth (1999) comments that the European words for 'hand' and 'first' are derived from the Latin word for 'finger' (Butterworth: 1999: 95). Shannon (1949) notes the base 2 is used in information theory and the resulting units may be called 'binary digits' or 'bits' as suggested by J. W. Tukey (Shannon and Weaver, 1949: 4). Solana-Ortega notes that the first name for such a 'binary digit' was 'bigit' or 'binit' from a 'binary unit' of information (Solana-Ortega, 2002: 465). Singh (1966) notes that 'bit' is a portmanteau of a 'binary unit' (Singh, 2002: 14).

The literature for this review represents the core of Algorithmic Information Theory/Kolmogorov Complexity and encompasses seminal, primary and secondary works in this subject area. A great deal of time and energy was taken to present the major focus of each article of information and draw a contextual story line from the material into a relevant whole.

Note: Current books on Algorithmic Information Theory and Kolmogorov Complexity, as of 2008, are as follows: Downey and Hirschfield (2007), Hutter (2005), Rissanen (2007), Salomon (2007). In searching the SAO/NASA ADS physics abstract the following citations were found: Benedetto, Caglioti, Loreto, and Pietronero (2002), Loreto and Puglisi (2003), Baronchelli, Caglioti, and Loreto (2005), Liu, Xiong, Wu, Wang, and Castleman (2001), Pappou and Tsangaris (1997), Lin, Athale, and Lee (1983), Tran (2007), Schmalz and Ritter (2005), Hayden, Jozsa and Winter (2002), Cathey (1984), Cherri (1996), YU, Liu, Mu and Yang (1998), Winograd and Nawab (1995), Li, Song, Wang, Jin and Zhang (2008), and Tice (2008).

The following papers are current as of 2008: Ferragina, Nitto and Venturini (2008), Calude and Zimand (2008), Dai and Milenkovic (2008), Tamaki (2008), Gagie (2006) Bol'shakov and Smirnov (2005) and Cooper and Lynch (1981).

Plate 4

4

The History of Algorithmic Information Theory

The foundations of Kolmogorov Complexity have come from von Mises' idea of random infinite sequences (Li and Vitanyi, 1993/1997: 89). As Li and Vitanyi state from their well researched text *An Introduction to Kolmogorov Complexity and Its Applications* (1993/1997):

> Komogorov complexity originated with the discovery of universal descriptions, and a recursively invariant approach to the concepts of complexity of description, randomness, and a priori probability. Historically, it is firmly rooted in R. von Mises's notion of random infinite sequences as described above. (Li and Vitanyi, 1993/1997: 89)

Li and Vitanyi consider Kurt Godel's 1936 paper 'On the length of proofs' that uses length as a measure of the complexity of proofs by proving that adding axioms to undecidable systems shortens the proofs of many theorems (Li and Vitanyi, 1993/1997: 89).

Kolmogorov Complexity and Algorithmic Information Theory both represent descriptional complexity, algorithmic information, and algorithmic probability (Li and Vitanyi, 1993/1997: 90). Both terms are be used synonymously in this thesis.

The actual inventors of Kolmogorov Complexity can be chronologically listed as: R.J. 'Ray' Solomonoff, of Cambridge, Massachussetts U.S.A., A.N. Kolmogorov, of Moscow, Russia, and G.J. 'Gregory' Chaitin, of New York City, U.S.A. (Li and Vitany, 1993/1997: 89). The time period for this 'discovery' was the 1960s with Solomonoff's papers of 1964, Kolmogorov's papers of 1965 and 1969, and Chaitin's papers of 1966 and 1969. Solomonoff had earlier papers on the subject, 1960, that were not widely cited until years later (Li and Vitanyi, 1993/1997: 89–90).

The question of why isn't Kolmogorov Complexity termed 'Solomonoff' Complexity, is due to the 'right of priority' in naming a discovery, is addressed by Li and Vitanyi in that it has become 'well entrenched' and 'commonly understood' (Li and Vitanyi, 1993/1997: 90). Both authors, Li and Vitanyi, suggest that Solomonoff be associated with the universal distribution and Kolmogorov with descriptional complexity (Li and Vitanyi, 1993/1997: 90).

A.N. Kolmogorov is pictured as a distinguished white haired gentleman in all of his 'official' public photographs while Ray Solomonoff's current picture, as seen on the internet site for IDSIA, is that of the classically wizened 'mad scientist' haired engineer that gives some merit to a recent article in *Science* magazine on young peoples perceptions of scientists (BBC News, 2006). Horgan in his book *The End of Science* (1996) describes Chaitin as being stout, bald, and boyish and dressed in 'neo-beatnik attire' (Horgan, 1996: 227). It sounds from Chaitin's attire that he is more Apple Computer than IBM when it comes to fashion. Chaitin is a member of the theoretical physics group at the IBM Thomas J. Watson Research Center in Yorktown Heights, New York (Chaitin, 1987: back cover).

Plate 5

5

The Sub-Maximal Measure of Kolmogorov Complexity

The notion of a 'sub-maximal' measure of Kolmogorov Complexity is from the fact that an algorithmically random string is defined as one of a near maximal information content. A maximal information content string is a string whose minimal program is about the same length as the string itself because the string lacks a significant internal pattern that would allow it to be compressed more completely (Bennett, 1989: 791).

The following is a traditional definition and example of a measure of randomness using Kolmogorov Complexity.

A Measure of Kolmogorov Complexity

Definition 1. Random or pattern less sequences of a given length are those that require the longest programs. Most of the binary sequences of length k require programs of about k length. These are random or pattern less sequences (Chaitin, 1970: 6).

Binary sequences that are shorter than the length of k are non-random sequences. The more it is possible to compress a binary sequence into a short program calculation, the less random a sequence (Chaitin, 1970: 6).

Notation 1. The following equation is used to prove the length of k of a program:

$L(M, S)$ less than or equal to $k + 1$ for all binary sequences S of length k,

where M = computer, S = sequence, k = length of sequence.

Lemma 1. *The following examples are taken from Chaitin (1970) to define examples of random and non-random binary bit strings (Chaitin, 1970: 6).*

Example:

Randomness
[A] 110010111110011001011110000010

Non-random
[B] 111111111111111111111111111111

Non-random
[C] 010010101010101010101010101010

Note: Chaitin (1970) has defined [A] to be more random, or more pattern less, that sequences [B] and [C] (Chaitin, 1970: 6).

Both sequences [B] and [C] can be 'more compressed' from their original lengths by multiplying 30×1 for [B] and 15×01 for [C] (Chaitin, 1970: 6). But [A] cannot be reduced from its original length, 30 bits long, because it is not 'compressible' to a more compressed, or 'shorter', definition from the original. This is the test for randomness and compressibility. If the string of binary bits cannot be compressed to less than it?s original size then it is random.

The Symbolic Space Multiplier Program (Tice, 2003: 60–61)

The following is a measure of randomness of Kolmogorov Complexity using the 'symbolic space multiplier program' that results in a sub-maximal measure of the traditional measure of randomness of Kolmogorov Complexity.

Definition 2. By introducing a specifically valued element into a binary system of the program of a sequence of binary bit strings, a new result for the definition of random and non-random binary bit strings produces a new measure of Kolmogorov Complexity. The introduction of a 'multiplying' arithmetic unit to a sequence of binary bit strings by way of a space between specific binary bits.

Notation 2.

1. The number before the space is the number to be multiplied.

2. The code bit number following the space is the multiplier.
3. Two spaces concludes further multiplication procedures and hence returns the computer to the next operation.
4. The multiplier is designated by a single or multiple character digit code.

Note: The example used in Tice (2000, 2003) of the 'symbolic space multiplier program' is using binary bits and a space function and is not to be considered a true ternary system.

Example:
Using Chiatin's example for a random binary bit string the 'symbolic space multiplier program' will be initiated (Chaitin, 1970: 6).

Randomness
[A] 1100101111100110010111110000010

Step One
The number before the space is the number to be multiplied.

Example:
[A] 110010[1] 001100101111[0] 110

Note: Bracketed symbols [ ] represent the symbol to be multiplied.

Step Two
The code bit number following the space is the multiplier.

Example:
[A] 1100101 {1}0011001011110 {1}10

Note: The parenthesis { } represents the multiplier. Step Four has the key code for {1} as representing the multiplier as 5.
    Hence {1} = 5 × [ ] = the original bit length.
    As the first set of five similar sequential symbols are 1's and the second set of five similar sequential symbols are 0's the arithmetic of $1 \times 5 = 11111$ and $0 \times 5 = 00000$ gives an accurate reproduction of the desired original bit lengths.

Step Three
Two spaces concludes further multiplication procedures and hence returns the computer to the next operation.

Example:
[A] 110010111110011001011110000110 (end of string; end of operation).

Step Four
The multiplier is designated by a single or multiple character digit code.

Example:
The multiplier in this operation is designated by the following single digit codes (Tice, 2003: 61).

Multiplier	Code
4	0
5	1

Notation 3. The following is the equation used to prove the length of k of a program:

$L(M, S)$ less than or equal to $k + 1$ for all binary sequences S of length k

Key: M = computer, S = sequence, K = length of sequence.

Proof. The following are results from using the 'symbolic space multiplier program'.

Original length of a random binary bit string of example [A] is 30 character bits (Chaitin, 1970: 6).

[A] 110010111110011001011110000010

After using the 'symbolic space multiplier program' on the example of a random sequence of binary bits [A] results in a character bit length of 22.

[A] 1100101 10011001011110 110

This qualifies for the [k] value for the length required to be less than the original length of a sequence of a binary bit string. Thus it satisfies the qualification as a 'patterned' or 'non-random' sequence of binary bits as defined by the equation for randomness for Kolmogorov Complexity. This results in a new measure of randomness for Kolmogorov Complexity as it is a 'sub-maximal' or reduced measure for what was suppose to be a 'random' sequence of a binary bit string.

Note: A interesting situation occurs with the 'symbolic space multiplier program' by including the 'reduction' or 'compression' of the similar sequential bits that are four bits long. Using the already compressed example [A]:

Example:
[A] 1100101 1001100101111 110

By using the 'symbolic space multiplier program' the following will result:

Step One

Example:
[A] 1100101 100110010[1]111 110

Step Two

Example:
[A] 1100101 100110010[1] {0} 110

Step Three

Example:
[A] 1100101 1001100101 0 110

Step Four

Key Code:

Multiplier	Code
4	0

Example:
[A] 1100101 1001100101 0 110

Result:
From the original 30 bit length and the introduction of the compression of two groups of five similar sequential bits that resulted in a 22 bit length, the compression of four similar sequential bits has actually reduced one bit length from the total bit length which is now 21 bits in length. This is important because it shows a potential level of saturation to this method of compression with the optimal compression, the greatest reduction of total bit length, being the two groups of five similar sequential bits from the original 30 bit length. In other words, a boundary maybe introduced to the 'sub-maximal measure of Kolmogorov Complexity' that seems to show a lower boundary limit to the 'symbolic space multiplier program'.

Plate 6

6

Binary, Ternary, and Quaternary Systems

A binary system is composed of two values while a ternary system is made up of three values and a quaternary system has four values. The binary system has the traditional role as representing a two value system and is listed as such in modern English dictionaries (Simpson and Weiner, 1989a). The ternary terminology has been taken from a dictionary source for validity as ternary has the English meaning of 'consisting, or composed of a set of three; threefold, triple; ternary' (Simpson and Weiner, 1989c). Quaternary is a four value system (Simpson and Weiner, 1989b).

A binary system is one that has only two values, usually 0 and 1, and in a perfect communication system, such as the telegraph system, i.e. Western Union, perfect transmissions, dot and dash, are the result of 'noiseless' systems with no 'entrope' to the signal source (Hankerson et al., 1998: 38) and (Endnote #15). This measure of self information is the 'bit' short for binary digit (Fano, 1961: 27). The antiquated terms such as 'nat' for 'natural unit' and 'Hartley' in honor of R.V. Hartley, a pioneer in communication theory, have been replaced by binary digit or 'bit' (Fano, 1961: 27 and 36). The term 'bit' was dubbed by J.W. Tukey (Millman, 1984: 387).

The Ternary system is composed of three symbols. One symbol more than the binary system. In the traditional alphabet symbol notation systems the symbol 0 signifies a quantity of zero and the symbol 1 signifies the quantity of one. In the binary system as devised by Shannon (1948) the symbols are just opposites of the other, 0 and 1 can represent anything as long as they do not represent the other as they are the contrastive features that give the binary value to the system. A classic Turing machine will have a binary system of symbols, usually 0 and 1, and that it will have an output of only 0 and 1, and unless an error occurs, such as a 'fuzzy' digit, two digits being produced at the same time, the resulting output will duplicate the input, that being the traditional binary 0 or 1 symbols (Hankerson, 1998: 37).

An interesting variation on this would be the addition of a mix of the 0 and 1 symbols, one symbol overlapping the other, to produce a hybride symbol that could be used as a 'trinary' system (Endnote #16). In the classic Turing machine model of computers it is not uncommon to see a blank space be incorporated into the binary system, a zero, a one, and a blank space [0], [1], and [ ] that in some respects make the blank space a 'pseudo-ternary' system to the binary system. This would only need a two value symbol type face to reproduce such symbols.

The quaternary system is one that has four symbols in the system. To use features that were proposed for a ternary system let the quaternary system have a zero, a one, a hybride of a zero and a 1, and a blank space resulting in [0], [1], [hybride 0 and 1] and [ ]. This would need only a two value symbol type face. This again may make it a pseudo-ternary system, because of the blank space, and even a pseudo-binary system, in that the blank space is ignored and the hybride 0 and 1 are only a 'net' result of the simultaneous reproduction of a 0 and 1 symbols, that will result in three symbols, does not change the nature of the type face in a classic Turing machine. Knuth (1998) makes the comment that a system called a 'quarter-imaginary' number system, analogous to a 'quaternary system', with the unusual feature that every complex number can be represented with the digits 0, 1, 2, and 3 without a sign (Knuth, 1998: 205; Knuth, 1960).

Four different sets make up the foundation of quantum logic operations: AND, OR, NOT, and COPY (Lloyd, 2006: 113). These properties were first ascribed by Fredkin and Toffoli to describe atomic collision's as the language of information processing, a quantum computer (Lloyd, 2006: 97).

Plate 7

7

Monochrome and Chromatic Symbols

Most binary systems do not usually value the color trait of the binary symbols, 0 and 1, in the information content of those binary symbols. Usually monochromatic, usually black, the two symbols. 0 and 1, have no color value added to them as an information unit. For the sake of novelty, the addition of color to the information content of a binary system would increase the amount of information carried by each binary symbol, 0 and 1, and the pattern of sequential binary sets would also have a corresponding color pattern value to the information corpus.

Also the amount of information per unit or bit is vital to a transmission system. Some bits of information, either a 0 or a 1, are singular in that no value is weighted to the fact that being a monochromatic entity is of no value to the amount, or 'content', of that unit or bit. Chromatic values increase the amount of information stored or carried in a unit or bit and can be based on variations of the color spectrum. Multiple units or bits, sequences of 0's and 1's, in both random and non-random patterns can form information patterns by corresponding color contrasts and similarities. These chromatic patterns or 'spectral grammars' are information features that add to a binary system without changing the binary digit fundamentals of the binary system.

Aspects of printing symbols is only touched on by Minsky (1967) but Brillouin (1956/1962) cedes a whole chapter to the issues arising from entropy and information in regards to writing, reading and printing of information (Minsky, 1967: 152; Brillouin, 1956/1962: 259–266).

While this thesis focuses on 2, 3 and 4 character types for symbols, the real limits to the types of characters used in modern methods of transcribing our thoughts to a permement media is that method of transcribing, the common keyboard. The modern 104 key QWERTY PC United States (English) keyboard is a development from the typewriter keyboard that employs extra keys for specific computer functions (Wikipedia, 2008, Keyboard-computing). The QWERTY keyboard was designed by Christopher Sholes in 1874 and stands for the top six characters of the keyboard (Wikipedia, 2008, QWERTY). The keyboard is the real limiter to the number and types of characters used to represent symbols in a systematic and logical fashion.

Plate 8

8

Aspects of Data Compression

In the field of information theory the standard data compression codes are those of Shannon, Fano, and Huffman (Hankerson, Harris and Johnson, 1998: 106–107). Both Shannon and Fano, now termed Shannon–Fano codes, are techniques for data compression using a prefix code based on a set of symbols and their estimated or measured probabilities. Because Huffman codes produce optimal prefix codes and are as computationally as simple as Shannon–Fano codes, Huffman codes have relegated the earlier codes of Shannon and Fano to historical footnotes. Arithmetic coding is also an optimal coding system (Ralston, Reilly and Hemmendinger, 2000: 493).

Li and Vitanyi state that language compression is closely related to the Kolmogorov complexity of the elements in the language (Li and Vitanyi, 1997: 477).

Minsky states that there is no finite-state multiplying machine which will work for arbitrarily large numbers (Minsky, 1967: 27; Eijck, 1994: 1244). If an example of the 'symbolic space multiplier program' is used to multiply an arbitrarily large number the finite-state nature of the program is observed while maintaining a functional multiplication program for an arbitrarily large number (Tice, 2003: 60–61).

Example:
The Symbolic Space Multiplier Program states (Tice, 2003: 60–61):

1. The number before a space is the number to be multiplied.
2. The code bit number following the space is the multiplier.
3. Two spaces concludes further multiplication procedures and hence returns the computer to the next operation.

4. The multiplier is designed by a single or multiple character digit code.

Hence the Symbolic Space Multiplier Program will function as a finite state machine that will multiply arbitrarily large numbers. This is due in large measure to the compression factor of using such a program.

An even greater level of compression, 'super compression', results when using a 'modified symbolic space multiplier program' when the first character symbol before a space functions as both the character symbol to be multiplied and as a key code representative of the multiplier with the space used to signify the character symbol to be multiplied.

Modified Symbolic Space Multiplier Program

1. The character symbol before the space is both the symbol to be multiped and the multiplier as directed by the key code guide.
2. The key code uses the character symbol to be multiplied to be the multipier. The character can represent only one value of multiplication although different character symbols can represent the same value of multiplication through out the string.
3. This system was designed to utilize multiple radix based character systems.

Example:
A binary bit string [A] of a character bit length of 20 bits.
[A] 11000111000001110011

Using the following key code

Key Code
$1 = \times 3$
$0 = \times 3$

The three similar sequences of binary bit characters results in a string length of binary bits, 1's and 0's, of 14 bits.
[A] 110 1 000001 0011

The 'symbolic space multiplier program' would have compressed the original 20 bit length to a character length of 14 bits as seen in example [B].

Example [B]:
[B] 11000111000001110011

Key Code
1 = 3
0 = 2

The first procedure is to reduce all similar characters of sequential threes.
[B] 110-11-1000001-10011

Resulting in a binary bit length of 14 characters.

Then compress all sequential similar character groups of twos.
[B] 1-00-1-01000001-10-01-0

Resulting in a binary bit length of 17 bits.

Of note is that the lower boundary of the 'symbolic space multiplier program' has been reached and a point of saturation has been reached and breeched. While still lower than the standard notion for randomness in a measure of Kolmogorov Complexity, the compression of only sequential groups of three character bits compressed to an optimal level of compression. For this study the optimal length of [B] will be used as an example.

   If a ternary or radix 3 based character system is used then the following result will occur.

Example:
[B] 11000111oooooo0001100

Using the 'modified symbolic space multiplier program' on example [B] the following will result:

Key Code
1 = ×3
0 = ×3
o = ×5

[B] 110-1-o-0-1100

Resulting in a character length of 10 trits.

Note: A trit is a ternary system term for three characters. Similar to bits for a binary bit system.

The ternary system was compressed by 50 percent. Revealing not only greater compression but also a more 'utilizable' character system than the traditional binary bit system.

If a radix 4 based character system, four separate character types or symbols, is used using example [C] and the 'modified symbolic space multiplier program' the following result will occur.

Example:
[C] 11000111oooooOOO11oo

Key Code
$0 = \times 3$
$1 = \times 3$
$o = \times 5$
$O = \times 3$

[C] 110-1-o-O-11oo

Resulting in a quaternary character length, radix 4 based character length, of 10. The quaternary based character length was compressed by 50 percent.

Again the quaternary based system proved to be more efficient and more 'utilizable' than the standard binary bit system with a compression ratio of 50 percent, same as the ternary based system.

The use of the 'modified' symbolic space multiplier program rather than the original 'symbolic space multiplier' system (Tice, 2003) produces not only a sub-maximal measure of Kolmogorov Complexity but compression rations of 2 to 1, 50 percent, in both a radix 3 and radix 4 based character systems. While I have focused on algorithmic information theory and Kolmogorov Complexity, the resulting compression of binary, ternary and quaternary based systems have direct practical applications to information theory and communication systems as a whole (Appendix A and Appendix B).

Plate 9

9

Conclusion

The standard definition of a measure of randomness to be found in a random string of binary bits has been examined using the 'symbolic space multiplier program' with the result a new measure to the notion of randomness of Kolmogorov Complexity. While this research was directed at random binary bit strings as defined by algorithmic information theory the developments beyond these parameters has lead to the introduction of both the radix 3 based and the radix 4 based character systems to both algorithmic information theory and later to information theory. The compression factor of almost one third to a random binary bit string using the 'symbolic space multiplier program' and the compression by half of the ternary and quaternary based systems to random strings using the 'modified symbolic space multiplier program' have deep theoretical and applied relevance beyond the fields of algorithmic information theory and information theory.

The sub-maximal measure of the randomness of Kolmogorov Complexity has fundamental implications that spread beyond algorithmic information theory in that the notion of 'information' is a measurable quantity and that the current thinking in the physical sciences has adopted fundamental aspects first developed by Shannon's information theory. With this in mind, the research from this thesis has explored a new standard of the notion of what makes up aspects of 'randomness' and sets a fundamental standard to the question of information as is currently defined in the literature. If 'information' is to be treated as a physical science, as it is currently done today, then the research found in this thesis is foundational to the notions of 'fundamental laws' that govern the universe. The physics of information as developed in this work are central to our understand of the known world and are more than an engineering or philosophical measures of thought. They are a measure of our world.

10

Summary

As can be seen from the use of the 'symbolic space multiplier program' on strings of binary bits, the compression factor used to define a level of randomness in a binary bit string is lowered from the standard model of Kolmogoroc Complexity allowing for a new measure of randomness in Algorithmic information Theory and Kolmogoroc Complexity. When both a radix 3 based character system, a ternary based system, and a radix 4 based system, a quaternary system, are introduced to the 'symbolic space multiplier program', a sub maximal measures of Kolmogoroc Complexity results that parallels those found using the binary bit strings. When both the ternary and quaternary based systems are used in the 'modified symbolic space multiplier program' considerable compression results with both the ternary and quaternary based systems achieving fifty percent compressions in their respective strings.

While I have focused exclusively on algorithmic information theory and Kolmogorov Complexity in this thesis, applications to both Shannon's information theory, and communication systems as a whole, are apparent (Shannon, 1948). Both Appendix A and Appendix B in this academic work address the questions raised in this thesis to information theory and communications systems in general.

Development of the Thesis

This thesis has the qualities of both a monograph on 'The History of Algorithmic Information Theory' and one on 'Data Compression and Optimal Coding Using Algorithmic Information Theory'. While I plan to rewrite this thesis and publish it as 'Algorithmic Information Theory: A Codex' the potential for future developments from this work is more than I had originally imagined, as I viewed it as a final statement on this field of study, while developing fundamental aspects to information theory, communications theory, computer science, mathematics, logic, engineering, philosophy and the growing interest of 'information' in physics.

Notes

1. In the field of computer science data is considered a coded representation of numbers, alphabetic characters, and special characters that are used in the operation of computation by a computer (Bitter, 1992: 223). Bitter makes the note that data is the plural of datum and in a grammatical sense should always be used with a plural verb, but common usage has it used with a singular verb (Bitter, 1992: 223).

2. An amusing side note to the concept of the 'meaning' of poetry may be gleamed from an old magazine article, 1962, that has a 'computer' programmed to produce 'beat' poetry (Horizon, 1962: 96–99). In this article a 'West-Coast' group of scientists programming a 'cool calculator', a computer, to 'create' novel lines of 'poetry' (Horizon, 1962: 98). They have named this computer A.B. for 'Auto-Beatnik' (Horizon, 1962: 98). One wonders about the 'value' of machine produced 'poetry', let alone the 'alternative', read 'beat' type, variety.

3. The Wikipedia Encyclopedia has the following names under the entry Kolmogorov Complexity: descriptive complexity, Kolmogorov–Chaitin complexity, stochastic complexity, algorithmic entropy, and program-size complexity (Wikipedia Encyclopedia, 'Kolmogorov Complexity', 1–7). Li and Vitanyi (1993/1997) have also included the following proper names for Kolmogorov Compexity: K-complexity, Kolmogorov–Chaitin randomness, algorithmic complexity, descriptional complexity, and minimum description length (Li and Vitanyi, 1993/1997: vi).

4. Not unlike some 'legacies' from the 'dark continent' (Jensen, 1963: 116–117).

5. Fuzzy logic is the development from the fuzzy set, introduced by L.A. Zadeh in 1965. The 'fuzzy set' represents a class who's characteristics have no sharp boundaries (Ralston, Reilly and Hemmindinger, 2000: 140). The transition from non-membership [0] and full membership [1] is gradual, rather than abrupt, with some elements being intermediate, those that are considered as 'marginal' or 'less acceptable', between either [0] and [1] (Ralston, Reilly, and Hemmindinger, 2000: 140).

6. Chomsky makes an interesting comment in Horgan (1996) in that he 'was almost totally incapable of learning languages' and that he was not even a 'professional linguist' (Horgan, 1996: 150). Two of Chomsky's 'famous' early works on linguistics are 'Syntactic Structures' (1957) and 'Aspects of the Theory of Syntax' (1965). Unfortunately, I have had to 'deal' with the 'social' aspects to Chomsky's 'revolution' in that I was always being harassed by either pro-Chomskian's, usually 'hippy types', and anti-Chomkians, usually military-industrial 'block-head' types, that raised my blood pressure and lowered my I.Q. by their constant and out of date, it was at the time the late 1980s through the mid 1990s, 1987–1994, 1960s 'culture wars'. These 'language wars' are examined in Harris (1993). I quickly made the move from linguistics to ESL, English as a Second Language, a field where the people are more contemporary and less 'crazed'. It was not much of a leap from Chomsky's early formal language work of the mid 1950s through the early 1960s to Algorithmic Information Theory because Chomsky was able to unite known logical systems and automata models in his abstract linguistics that carried over into programming languages (Millman, 1984: 387). The Harvard psychologist Steven Pinker is probably the most popular of those who have 'translated' Chomsky's language theories with such popular works as *The Language Instinct* (New York: W. Morrow and Company. 1994). Pinker has written an interesting paper (2006) that seems to develop a union between the sciences and humanities that I give some credence to in that the fundamental ideas found in my work come from areas found from both disciplines (Pinker, S. 'The humanities and human nature'. *Skeptical Inquirer*, Volume 30, Issue 6, November/December 2006, pp. 23–28).

7. Chaitin's 1975 *Scientific American* article was an unintentional vehicle for the development of a sub-maximal measure of Kolmogorov complexity in that when I first read the article I was in a state of

agitation and I looked at the 'nonrandom' series of '1's' and '0s' and I just decided to 'compress' the string of these '1's' and '0s' into concatenated groups of similar sequences with a 'space' to note an action of arithmetic, to 'multiply', by a specific number to duplicate the required number of '1's' or '0s'. The agitation was a by-product of a 'dyspepsia' induced by an 'indigestible' meal. In other words my gut feeling was really a GUT feeling. The properties of a 'visual' intelligence is addressed in Kemp's article from a book of the same topic (Kemp, 2006: 48–49). I find that my 'visual' perception of what was 'laborious' or 'redundant' in the Chaitin (1975) article of the segments of string of '1's' and '0's' in sequential groups was the vital key to the development of a new measure of Kolmogorov complexity. It seems I had arrived at the answer before I had the question.

8. For a concise biography of Shannon see the 'Biography' section of Shannon's collected papers (1993) edited by Sloane and Wyner. Shannon relieved his Master's of Science degree in electrical engineering, awarded in 1937, at MIT that was titled 'A symbolic analysis of relay and switching circuits' and was heralded as 'one of the most important master's thesis ever written' by H.H. Goldstine, because Shannon had studied relay and switching circuits and incorporated symbolic logic and Boolean algebra into a two value, binary, system and that it could be used for that analysis and synthesis of such processes (Sloane and Wyner, 1993: xi–xii, 469–495). Shannon's doctorial work in algebra, at MIT, for population genetics, awarded in 1940, was titled 'An algebra for theoretical genetics' and has been something of an enigma to geneticists in that it is 'entirely unknown to contemporary population geneticists' (Sloane and Wyner, 1993: 921). Weil (2003) notes that Shannon had conceptualized his dissertation while working with Barbara Burks, a geneticist, at the Cold Springs Harbor laboratory during the summer of 1939 (Weil, 2003: 493). Weil (2003) states that Shannon's Doctorial work, awarded in 1940, greatly aided in the organization of genetics (Weil, 2003: 493).

9. Bell laboratories has an almost 'mythical' legacy in being an institution that supported great scientific discoveries. While it's parent company, AT&T, has been split up, down sized and sold to other companies, its current incantation has it as at&t; in 'small' letters, Bell labs is still one of the largest and most productive private laboratories in the

country (Stokes, 2006: 3; Hochfelder, 2002; Cauley, 2005; Henck and Strassburg, 1988). Bell laboratories also has 'Bell Labs' in China, India and Ireland. It seems that research has become a 'global' phenomena beyond America's frontier. I am sure that if Horace Greeley were alive today he would have said 'Go East young man' instead. Ernst (2006) notes the rapid development of research laboratories in Asia as 'innovation' goes 'offshore' (Ernst, 2006: 29–33).

10. Gell-Mann addresses complexity in his book *The Quark and the Jaguar* (1994). The biography of Gell-Mann was edited by George Johnson and seems to portray Gell-Mann as a bi-polar, trinket stealing scientist with a bad case of writers block (Johnson, 2000). The book even tells of how the other publishers termed Gell-Mann's book *The Quark and the Jaguar* (1994) as 'The Jerk and the Quagmire' because of the delays in submitting the manuscript to the publisher (Johnson, 2000: 344). Gell-Mann even failed to submit his official 1969 Nobel Prize Lecture for the annual celebratory volume (Johnson, 2000: 10). A more honest account of Gell-Mann can be found in a scientific volume that details his life in science (McMurray, 1995). In this biographical sketch Gell-Mann's interest in linguistics, especially 'new' word formations, is given light, as well as his wide area of interests, including educational reform (McMurray, 1995: 745).

11. Shannon makes the comment to Anthony Liversidge, in an interview, that Wiener didn't have much to do with information theory and that he, Wiener, was not a big influence on Shannon's ideas about information as entropy (Sloane and Wyner, 1993: xxvii). Shannon had reviewed Wiener's book *Cybernetics* (1948) for the *Proceedings of the Institute of Radio Engineers* (1949) and finds the book an 'excellent introduction' into the filed of communication theory (Sloane and Wyner, 1993: 872–873). Shannon makes this claim even after citing 'numerous misprints' and a 'few errors of over-simplification' in Wiener's book (Sloane and Wyner, 1993: 872). This seems to be a common feature in technical book reviews as Jurgen Schmidhuber reviews Seth Lloyd's *Programming the Universe* (New York: Alfred A. Knopf, 2006) for the *American Scientist*, July–August edition 2006, that attack's most of the book with such comments as 'Some of Lloyd's statements reflect a certain naivete about some topics in computer science' and then points to a fundamental error in describing the Church–Turing Hypothesis

(Schmidhuber, 2006: 364). Schmidhuber concludes his review with the comment: 'Despite my few quibbles, I recommend this well written book without hesitation to anybody interested in an overview of basic ideas in the field. I intend to buy a few copies as presents for my friends' (Schmidhuber, 2006: 365). If he, Schmidhuber, writes book reviews like this one it is little wonder he has any friends at all. Why, after trashing a book, would one endorse that book? The few 'quibbles' about the book run the entire length of the two page book review. The point is if you give a negative review of a book maybe you shouldn't endorse that book.

12. It is interesting to note that the idea of 'newer' in terms of technology is not always better. The Apple iPod device that is hailed as the 'next big thing' in consumer electronics has the novel feature of being played at potentially damaging sound levels, 115 decibel level, but still manages to be considered 'The Perfect Thing' by Steven Levy (2006). Even the first portable transistor radio, The Regency TR 1 (1954), had a volume control (Riordan and Hoddeson, 1997: 212). Apple's products were even developed by other company's technology, most notoriously, PARC Xerox as depicted in Hiltzik's Dealers of Lightening (Hiltzik, 1999). Perhaps a book should be written on Apple's early history with the title *Stealer's of Lightening*. My first encounter with Silicon Valley 'truths' came in 1973 when I was in Junior High School in Palo Alto, California and my friend had told me his father, Nolan Bushnell, had 'invented' the video game called 'pong'. It would be years later that I would find out, and so it would seem the world, that Mr. Bushnell did not invent 'pong' but rather a Mr. Ralph Baer in 1966 who later joined with Magnavox with the product coming onto the market in the early 1970s. Other Silicon Valley 'legends' can be found in Stoll's (1995) *Silicon Snake Oil*.

13. The field of artificial intelligence celebrated its 50th founding conference at Dartmouth College on July 13–15, 2006 with John McCarthy, Marvin Minsky, and Ray Solomonoff attending as original 1956 conference participants. McCarthy, then on Dartmouth College's mathematics faculty, coined the term 'artificial intelligence' to emphasize the Project's focus in this 1956 conference.

14. Words have the duality to be both clear and ambiguous. Words such as 'information', 'knowledge' and 'intelligence' have a social power well beyond dictionary meanings. Information has had a long standing,

common-sense meaning before Shannon used it to describe 'entropy' in a communications system (Roszak, 1986: 13). Roszak notes the popular use of 'information' and 'knowledge' as if they were synonymous, especially as it applies to the notions of the mechanics of social forces; the 'information age' (Roszak, 1986: 22). This age of information is due in no small part to the advent of microelectronics that have allowed a 'closed' system for utilizing information in the form a single medium of electronic signals (Barron and Curnow, 1979: 27). Hence, the bases for an information society is one that has a robust and reliable communications system (Barron and Curnow, 1979: 31). One can see the same techniques being applied to sell the 'internet' especially as a 'learning' tool for educating children (Tapscott, 1998: 127–157; Talbott, 1995: 143–149). The generic type of marketing 'hype' book can be seen in Tapscott (1998) that can be balanced by the more 'dystopian' work such as Talbot (1995). Intelligence is also a misused word that beyond the statistical measure used by psychologists has little or no reason for use especially when trying to define 'athletic' traits that seem more appropriate to a record book than a text book (Roszak, 1986: 13).

15. The nature of a ternary system is common in nature as a 'system of three's' seems to have a 'natural' robustness over other number types (Thompson, 1961: 260). In the study of languages the structure of the Arabic language is best known for its interdigitated morphology (Asher and Simpson, 1994: 193). What is meant by 'interdigitated' morphology is that the basic morphological unit in word building is a large set of mostly triconsonant roots with a small corpus of fixed consonant-vowel patterns that are applied to these roots to generate various categories of verb and noun stems (Asher and Simpson, 1994: 193). The most common triconsonant root in modern literary Arabic is KTB, having to do with writing, of this algebraic-looking Semitic language grammar (Kaye, 1990: 665–666).

16. Al-Khwarizmi, Muhammad ibn Musa al-Khwarizmi, was a Persian polymath, mathematics, astronomy, astrology, and geometer, and author of *The Compendious Book on Calculation by Completion and Balancing* (820 A.D.) (Wikipedia, 'The Compendious Book?', 2006: 1). The work seems to have connections to Indian and Hebrew texts as there are no citations and is considered by scholars to be a compilation of knowledge from the Muslim world (Wikipedia, 'The

Compendious Book?', 2006: 1). An interesting event occurred when I tried to type Al-Khwarizmi's name in the reference section of this thesis and was changed by the spell checking system on my computer to 'Al-Charisma'. This effect of unintentional changing of words due to automatic spell checkers is known as the 'Cupertino Effect' (Biggar, 2006: 1). Biggar (2006) notes that automatic computer spellcheckers tend to replace 'cooperation' with 'Cupertino', a city just north of San Jose, California, with a new word coined for such a process of words erroneously changed and inserted into documents (Bigger, 2006: 1). Such words as 'prosciutto' are mistaken for 'prostitute' and 'identified' for 'denitrified' and this 'Cupertino Effect' has its origins going back to 1989, with documents in 2000 for the European Union, EU, being rift with 'Cupertino' instead of 'cooperation' (Bigger, 2006: 1).

17. In 2004 I moved my company, Advanced Human Design from Cupertino, California, established in 1992, to the Central Valley of Northern California. This move was in part due to the 'quality of life' issues that had developed in the Bay Area over the last decade (Seyfer, 2006: 1–2).[1]

---

[1] Seyfer, J. (2006) Bay area brain drain. Silicon Valley.com., Friday, March 24, 2006., pp. 1–2. Web address: http://www.siliconvalley.com/mld/siliconvalley/14177694.htm?template=contentModules/.

Appendix A
A Ternary Based System for Information Theory*

Abstract

A radix 3 based system of characters, or symbols, composed of three separate symbols will be examined to prove that it is a more efficient, robust and 'utilizable' than a binary bit system that is currently used today for information theory.

Introduction

Information theory was developed by Shannon in 1948 with the fundamental unit of 'information' based on a radix 2 system composed of two symbols, a 1 and a 0, each having no semantic value other than being the opposite of the other symbolically (Shannon, 1948). Richards (1955) has noted that the radix three, a ternary system, as the most efficient base, more so than the radix two and radix four base systems (Richards, 1955: 8–9). Because the radix 3 based system, or ternary system, is composed on 3 separate characters, the ternary system will be examined to show it to be more efficient, robust and utilizable than the binary system. A proofing operation consisting of a 'modified symbolic space multiplier program' will provide a compression value that will be lower than the Kolmogorov Complexity value found in algorithmic information theory that used a binary base of random strings (Tice, 2003).

---

*Technical Paper, *Advanced Human Design*, Volume 1, Number 1, December 2006, reprinted here with kind permission from the publisher.

Part I

The 'modified symbolic space multiplier program' is a development from the 'symbolic space multiplier program' developed by Tice (2003) using a random string of binary bits.

The Modified Symbolic Space Multiplier Program

1. The character symbol before the space is both the symbol to be multiplied and the multiplier as directed by the key code guide.
2. The key code uses the character symbol to be multiplied to be the multiplier. The character can represent only one value of multiplication although different character symbols can represent the same value of multiplication through out the string.
3. The system was designed to utilize multiple radix based character systems.

An example [A] of a random string of ternary characters 20 trits in length.

   Note: A trit is a ternary system term for three characters. Similar to bits for a binary bit system.

Example [A]:

[A] 11000111oooooo0011oo

Using the Key Code on example [A] the following will result as seen in example [B]:

Key Code

$1 = \times 3$

$0 = \times 3$

$o = \times 5$

Example [B]:

[B] 110-1-o-0-11oo

Resulting in a character length of 10 trits.

Part II

The ternary system was compressed by 50 percent from the original random 20 character trit string. This is a more compressed value than Kolmogorov Complexity has for a random string as well as being a more compressible form than the traditional random binary bit string (Tice, 2003: 64). The ternary system has shown a more efficient level of compression, a greater flexibility and utilization than traditional binary bits and is a more robust system because of these features that seem inherent in the radix 3 based system.

Summary

A ternary based system has been shown to be more compressible than a binary based system and has important aspects to algorithmic information theory regarding Kolmogorov Complexity. These qualities of compression have important developments in the areas of transmission and storage with respect to information theory as well as to a fundamental understanding of the laws that governing our world.

References

Richards, R.K. (1955) *Arithmetic Operations in Digital Computers.* New York: D. Van Nostrand Company, Inc.

Shannon, C.E. (1948) A mathematical theory of communication. *Bell System Technical Journal,* July & October 1948. Reprinted in Slepian, D. (1973) *Key Papers in The Development of Information Theory.* New York: IEEE Press, pp. 5–18 and 19–29.

Tice, B.S. (2003) *Two Models of Information.* Bloomington: 1st Books.

Appendix B
A Radix 4 Based System for Communications Theory*

Abstract

The introduction of a radix 4 based system composed of four separate characters that forms a more efficient, robust and utilizable system than traditional binary bits. This has important applications to communication theory.

Introduction

Brilliouin noted 50 years ago that the science of information has found a diverse application to telecommunications, computing, pure physics and to the fundamental process of scientific observations (Brillouin, 1956/1962: 1). Information theory as developed by Shannon in 1948 was based on a binary bit system that has served information theory, computer science and communication theory to this date (Shannon, 1948). The object of this paper is the introduction of a quaternary based system that is composed of four different characters, or symbols, that perform in a more efficient, robust and utilizable manner than the traditional binary system. This paper will introduce a proofing procedure termed a 'modified symbolic space multiplier program' that will present compression ratios that exceed the traditional measure of the Kolmogorov Complexity level found in algorithmic information theory. The foundational work for this compression factor can be found in Tice (2003) and is a development originally found using a random binary bit string.

*Technical Paper, *Advanced Human Design*, Volume 1, Number 2, December 2006, reprinted here with kind permission from the publisher.

Part I

The 'modified symbolic space multiplier program' was a development from the 'symbolic space multiplier program' developed by Tice (2003) using random binary bit strings.

The Modified Symbolic Space Multiplier Program

1. The character symbol before the space is both the symbol to be multiplied and the multiplier as directed by the key code guide.
2. The key code uses the character symbol to be multiplied to be the multiplier. The character can represent only one value of multiplcation although different character symbols can represent the same value of multiplication through out the string.
3. The system was designed to utilize multiple radix based character systems.

A random string of a radix 4 based character system, four separate character types or symbols, is used in example [A].

Example [A]

[A] 110001110000000001100

Using the key code guide the following will result in example [B]:

Key Code
$0 = \times 3$ $1 = \times 3$ $o = \times 5$ $O = \times 3$

Example [B]:

[B] 110-1-o- O-1loo

Resulting in a quaternary character length, a radix 4 based character length, of 10. This compression of a random set of a quaternary string of characters is reduced by 50 percent.

Part II

Because compression of a string of characters is fundamental to the optimal operation of communication transmissions and storage, the functional aspects of this compression ration to reduce the size of a string of symbols points to strengths of such a 4 character, or symbol, system. A quaternary based system becomes more efficient than a binary bit system because it can be compressed to a greater degree and provides a more robust system for information in that system (Tice, 2003: 64). In effect, the radix 4 based system becomes more 'utilizable' because of its ability to compress to a greater degree than a binary system and provides applicable standards to communication systems.

Summary

The compression ratio for a quaternary based system is superior to a binary system and has influential theoretical and applied aspects beyond algorithmic information theory to that of information theory and to communication theory as a whole.

References

Brilluoin, L. (1956/1962) *Science and Information Theory*. New York: Academic Press, Inc.

Shannon, C.E. (1948) A mathematical theory of communication. *Bell System Technical Journal*, July & October, 1948. Reprinted in Slepian, D. (1973) *Key Papers in the Development of Information Theory*. New York: IEEE, pp. 5–18 and 19–29.

Tice, B.S. (2003) *Two Models of Information*. Bloomington: 1st Books.

Appendix C
The Use of a Radix 5 Base for Transmission and Storage of Information*

Abstract

The radix 5 based system employs five separate characters that have no semantic meaning except not representing the other characters. Traditional literature has a random string of binary sequential characters as being 'less
patterned' than non-random sequential strings. A non-random string of characters will be able to compress, were as a random string of characters will
not be able to compress. This study has found that a randix 5 based character
length allows for equal compression of random and non-random sequential
strings. This has important aspects to information transmission and storage.

Keywords: Radix 5, Information Theory, Algorithmic Information Theory,
Transmission, Storage, Communication Theory.

Introduction

As communications handle an ever-growing amount of information for transmission and storage, the very real need for an upgrade in the fundamental
structure of such a system has come to light. As the very bases of coding
is compression, the greater the amount of information compressed, the more
efficient the system. The earliest calculating machine was the human hand,
its five digits representing a natural symmetry found, with frequency, in the
organic world [1, 2]. A radix 5 based system, also known as a quinary numeral
system, is composed of five separate characters that have no meaning apart

---

*Paper prepared for the Photonics West Conference in San Jose, California, Wednesday,
January 23, 2008, *Proceedings of the SPIE*, Volume 6896, pp. 68961H–68961H-7 (2008).

from the fact that each character is different from the other characters. This is a development from the binary system used in Shannon's information theory (1948) [3].

Part I

The radix 5 base is not the traditional binary based error-detection and error-correction codes that are also known as 'prefix codes' that use a 5-bit length for decimal coding [4]. A radix 5 base is composed of five separate symbols with each an individual character with no semantic meaning. A random string of symbols has the quality of being 'less patterned' than a non-random string of symbols. Traditional literature on the subject of compression, the ability for a string to reduce in size while retaining 'information' about its original character size, states that a non-random string of characters will be able to compress, whereas the random string of characters will not compress [5].

Part II

The following examples will use the following symbols for a radix 5 based system of characters [Example A].

Example A
o
O
Q
1
I

The following is an example of compression of a random and non-random radix 5 based system. A non-random string of radix 5 based characters with a total 15 character length [Group A].

Group A: oooOOOQQQ111III

A random string of a radix 5 based characters with a total of 15 character length [Group B].

Group B: oooOOQQQQ11IIIII

If a compression program were to be used on Group A and Group B that consisted of underlining the first individual character of a similar group of sequential characters, moving towards the right, on the string and multiplying it by a formalized system of artithmetic as found in a key, see Key Code A and Key Code B, with the compression of Group A and Group B as the final result.

Key Code A (for Group A)

o = ×3
O = ×3
Q = ×3
1 = ×3
I = ×3

Group A: oOQ1I

Resulting in a 5 character length for Group A.

Key Code (for Group B)

o = ×3
O = ×2
Q = ×4
1 = ×2
I = ×4

Group B: oOQ1I

Resulting in a 5 character length for Group B.

   Both Group A (non-random) and Group B (random) have the same compression values, each group resulted in a compression value of 1/3 the total pre-compression, original, state. This contrasts traditional notions of random and non-random strings [5]. These findings are similar to Tice (2003) and have applications to both Algorithmic Information Theory and Information Theory [6].

Some other examples using Example A Radix 5 characters [oOQ1I] to test random and non-random sequential strings.

The following is a non-random string of radix 5 based characters with a total of 15 character length [Group A].

Group A: oooOOOQQQ111III

A random string of radix 5 based characters with a total of 15 character length [Group C].

Group C: oooooOQQQQQ1I

If a compression program were to be used on Group A and Group C that consists of underlining the first individual character of a similar group of sequential characters, moving towards the right, on the string and multiplying it by a formalized system of arithmetic as found in a key, see Key Code A and Key Code C, with the compression of Group A and Group C as the final result.

Key Code C (for Group A)

o $= \times3$
O $= \times3$
Q $= \times3$
1 $= \times3$
I $= \times3$

Group A: oOQ1I

Resulting in a 5 character length for Group A.

Key Code C (for Group C)

o $= \times5$
O $= \times1$
Q $= \times5$
1 $= \times1$
I $= \times1$

Group C: oOQ1I

Resulting in a 5 character length for Group C.

This example has Group A as a non-random string and Group D as a random string using radix 5 characters for a total 15 character length.

A non-random string of radix 5 characters with a 15 character length (Group A).

Group A: oooOOOQQQ111III

A random string of a radix 5 based characters with a total of 15 character length (Group D).

Group D: oOOOOQQ1111III

If a compression program were to be used on Group A and Group D that consisted of underlining the first individual character of a similar group of sequential characters, moving towards the right, on the string and multiplying it by a formalized system of arithmetic as found in a key, see Key Code A and Key Code D, with the compression of Group A and Group D as the final result.

Key Code A (for Group A)

o $= \times 3$
O $= \times 3$
Q $= \times 3$
1 $= \times 3$
I $= \times 3$

Group A: oOQ1I

Resulting in a 5 character length for Group A.

Key Code D (for Group D)

o $= \times 1$
O $= \times 4$
Q $= \times 1$
1 $= \times 4$
I $= \times 4$

Group D: oOQ1I

Resulting in a 5 character length for Group D.

As a final example Group A is a non-random sequential string and Group E a random sequential string using radix 5 based characters for a total of 15 character length.

A non-random string of radix 5 based characters with a 15 character length (Group A).

Group A: oooOOOQQQ111III

A random string of radix 5 based characters with a total of 15 character length (Group E).

Group E: ooOOOOQQQQ111II

If a compression were to be used on Group A and Group E that consisted of underlining the first individual character of a similar group of sequential characters, moving to the right, on the string and multiplying it by a formal-ized system of arithmetic as found in a key, see Key Code A and Key Code E, with the compression of Group A and Group A as the final result.

Key Code A (for Group A)

$o = \times 3$
$O = \times 3$
$Q = \times 3$
$1 = \times 3$
$I = \times 3$

Group A: oOQ1I

Resulting in a 5 character length for Group A.

Key Code E (for Group E)

$o = \times 2$
$O = \times 4$
$Q = \times 4$

$1 = \times 3$
$I = \times 2$

Group E: oOQ1I

Resulting in a 5 character length for Group E.

Again, these examples conflict with traditional motions of random and non-random sequential strings in that the compression ratio is one third that of the original character number length for both the random and non-random sequential strings using a radix 5 based system.

Part III

Traditional information based systems use a binary based system represented by either a 1 or a 0. First developed by Claude Shannon in 1948 and termed 'information theory', this fundamental unit has become the backbone of our information age. One important aspect to information theory is that of data compression, the removal of redundant features in a message that can reduce the overall size of a message [7]. With the substantial compression values found in using a radix 5 based system it seems a new paradigm has arrived to carry the future of information.

Information technology has been the major drive of the economic growth in the past decade adding $2 trillion a year to the economy [8]. This growth needs to be sustained in order for new jobs and the economy to maintain a high standard of living. Only by considering alternative developments to existing models of technology, can the future of the economy develop and continue at a successful level of growth.

The internet was an outgrowth of the cold war as a government-sponsored project to develop communications network that was decentralized [9]. Today the internet is the major highway of global information with search engine technology rapidly taking center stage on both universities research departments as well as the Dow Jones index. The need to handle this vast and ever growing amount of information will need a fundamental change to the very nature of the structure of our information systems. It is clear that any new developments to deal with more and more information must begin at the fundamental level.

With a radix 5 base that has been proved to have the compression ratio similar in both random and non-random states, the question of usage as a medium for transmission and storage of information becomes paramount.

With an ever increasing need for transmission and storage in the areas of telecommunications and computer science, the viability of a new system at the fundamental level of communication theory that is both robust and diverse enough to allow for future growth beyond the binary based system in use today.

Summary

This paper has shown that a radix 5 based system has profound properties of compression that are well beyond those found in binary systems using sequential strings of a random and non-random types. These compression values have strong potential applications to information theory and communication theory as a whole.

While the identical compression values for random and non-random radix 5 based strings is a result of this paper, the application of this theory to communication theory cannot be understated. It has been shown that a radix based 5 based system has a compression factor that makes it an ideal functional standard for future information systems, particularly in the fields of telecommunications and computer science.

References

[1] G. Ifrah, *The Universal History of Numbers*, John Wiley & Sons, New York, 2000, p. 47.

[2] H. Weyl, Symmetry, in *The World of Mathematics*, J.R. Newman (Ed.), Simon and Schuster, New York, pp. 671–724 (p. 710).

[3] C.E. Shannon, A mathematical theory of communication, *Bell Sys. Tech. J.* 27, 1948, 379–423 & 623–656.

[4] R.K. Richards, *Arithmetic Operations in Digital Computers*, D. Van Nostrand Company, Princeton, 1955, p. 184.

[5] S. Kotz and N.I. Johnson, *Encyclopedia of Statistical Sciences*, John Wiley & Sons, New York, 1982, p. 39.

[6] B.S. Tice, *Two Models of Information*, 1st Books Publishers, Bloomington, 2003.

[7] B. Gates, N. Myhrvold and P. Rinearson, *The Road Ahead*, Viking, New York, p. 30.

[8] F. Davis, Impact of information technology touted, *Silicon Valley.com*, March 14, 2007, p. 1.

[9] J.E. Nuecherlein and P.J. Weiser, *Digital Crossroads*, The MIT Press, Cambridge, 2005, p. 129.

Appendix D
In Praise of Paperclip Physics*

My experience with Kolmogorov complexity began in 1998 while doing research for my company Advanced Human Design, that was then located in Cupertino, California. Advanced Human Design is a boutique sized research and development company that focuses on telecommunications and medical information sciences. The company's motto is 'Tomorrow's future today', but it should have been 'Yesterday's future tomorrow' as almost all of my work revolves around outdated and arcane subject matter that seems to work despite time and the progress of technology. One of the more interesting aspects of my work is that it involves paper and pencil research that seems bucolic to the big science of the late 20th and early 21st century world. Thus the term 'paperclip physics' in the title of this essay as it is both inexpensive and involves only myself.

In reading histories of the quantum sciences 'golden years' (1900 through 1930), I find a similar feeling of discovery that is not shackled by big budgets or armies of researchers. Gino Segre notes in his book *Faust in Copenhagen* (2007) that:

> Those discoveries of 1932 sometimes called the Miracle Year of experimental physics, also shifted the emphasis in physics from theory to experimental, from research done with a pencil and paper to research done with sophisticated tools in a laboratory. (Segre, 2007: 7)

A similar atmosphere is gleamed from reading 'Uncertainty' by David Lindley (2007) also about the development of quantum physics in the early years of the 20th century. Fueled by cafene from coffee and allowed time to think

---

*An essay prepared for *Advanced Human Design*.

97

in the modern day coffee houses, I was able to develop the necessary insights to develop concepts for Kolmogorov complexity as it related to algorithmic information theory and information theory.[1]

As this is 2008 and the 10th anniversary of my interest in Kolmogorov complexity, some points need to be made regarding this research. I copyrighted a manuscript in 2000 that has a chapter dealing with a sub-maximal value of Kolmogorov complexity as it relates to sequential binary strings. I published my Ph.D. dissertation as 'Two Models of Information', essentially a chapter from the copyrighted 2000 manuscript, in 2003 (Bloomington: AuthorHouse) and I copyrighted two papers dealing with radix 3 and radix 4 bases using Kolmogorov complexity in 2006 as well as copyrighting the first edition of this dissertation in 2006. A radix 5 based system dealing with Kolmogorov complexity was copyrighted in 2007. The results of this research have provided a 'fundamental' aspect to the known limits of Kolmogorov complexity that has both strong theoretical and applied applications in both algorithmic information theory and information theory. As it relates to the broad field of telecommunications theory, the very nature of the substantial compression values obtained in this research point to applications in the fundamental structure of information, mainly the archane binary bit system. The promise of quantum information theory is still a promise and new ideas and technologies are needed to get the market moving (Venema, 2007: 175). One last comment, the 2007 Nobel Prize for Physics was awarded for the discovery of the effect of the giant magnetoresistance (GMR) that allowed for compression of information on hard disc drives for the electronics industry (Brumfiel, 2007, 643).

The discovery of substantial compression in random sequences of strings that result in a 'sub-maximal measure of Kolmogorov complexity' have a direct effect on applied aspects of the telecommunications industry and communication theory as a whole and has a greater net effect on the industry as a whole than previous discoveries. This research seems to be the future of information, as we know it.

---

[1] An interesting book by Brian Cowan titled *The Social Life of Coffee: The Emergence of the British Coffeehouse* (New Haven: Yale University Press, 2005) gives a history to the 'coffeehouse' experience.

References

Brumfiel, Geoff (2007) The physics prize inside the iPod. *Nature*, Volume 449, Issue 7163, October 11, 2007, p. 643.

Lindley, David (2007) *Uncertainty: Einstein, Heisenberg, Bohr, and the Struggle for the Soul of Science*. New York: Doubleday.

Segre, Gino (2007) *Faust in Copenhagen: A Struggle for the Soul of Physics*. New York: Viking.

Venema, Liesbeth (2007) Reality check. *Nature*, Volume 450, November 8, 2007, p. 175.

Appendix E
A Comparison of a Radix 2 and a
Radix 5 Based System*

Abstract

A radix 2 based system is composed of two separate character types that have no meaning except not representing the other character type as defined by Shannon in 1948. The radix 5 based system employs five separate characters that have no semantic meaning except not representing the other characters. Traditional literature has a random string of binary sequential characters as being "less patterned" than non-random sequential strings. A non-random string of characters will be able to compress, whereas a random string of characters will not be able to compress. This study has found that a radix 5 based character length allows for equal compression of random and non-random sequential strings. This has important aspects to information transmission and storage.

Keywords: Radix 5, Information Theory, Algorithmic Information Theory, Transmission, Storage, Communication Theory..

Introduction

As communications handle an ever-growing amount of information for transmission and storage, the very real need for an upgrade in the fundamental structure of such a system has come to light. As the very bases of coding is compression, the greater the amount of information compressed, the more efficient the system. The earliest calculating machine was the human hand,

---

*A poster presented at the SPIE Symposium on Optical Engineering and Applications, held in San Diego, California, August 10–14, 2008.

its five digits representing a natural symmetry found, with frequency, in the organic world [1, 2]. A radix 5 based system, also known as a quinary numeral system, is composed of five separate characters that have no meaning apart from the fact that each character is different than the other characters. This is a development from the binary system used in Shannon's information theory (1948) [3].

The Radix 2 Based System

The radix 2 based system is a two character system that has no semantic meaning except not representing the other character type. The traditional 1 and 0 will be used in this paper.

The following is an example of compression of a random and non-random radix 2 based system. A non-random sequential string of characters will have a total length of 15 characters as seen in Group A.

Group A: 111000111000111

A random sequential string of characters will have a total length of 15 characters as seen in Group B:

Group B: 110000111110111

If a compression program were to be used on Group A and Group B that consisted of underlining the first individual character of a similar group of sequential characters, moving towards the right, on the string and multiplying it by a formalized system of arithmetic as found in a key, see Key Code 1 and Key Code 2, with the compression of Group A and Group B as a final result.

Key Code 1 (for Group A)

$1 = \times 3$
$0 = \times 3$

Group A: 10101

Resulting in a compressed state of 5 characters for Group A.

Key Code 2 (for Group B)

1 = ×5
0 = ×4

Group B: 11<u>01</u>0111

Resulting in a compressed state of 8 characters for Group B.

Compression values of the non-random binary sequential string are one third the original 15 character length and the random binary sequential string are almost half of the original 15 character length.

Part I

The radix 5 base is not the traditional binary based error-detection and error-correcting codes that are also known as 'prefix' codes that use a 5-bit length for decimal coding [4]. A radix 5 base is composed of five separate symbols with each an individual character with no semantic meaning. A random string of symbols has the quality of being 'less patterned' than a non-random string of symbols. Traditional literature on the subject of compression, the ability of a string to reduce in size while retaining 'information' about its original character size, states that a non-random string of characters will be able to compress, whereas the random string of characters will not compress [5].

Part II

The following examples will use the following symbols for a radix 5 based system of characters [Example A]

Example A
o
O
Q
1
I

The following is an example of compression of a random and non-random radix 5 based system. A non-random string of radix 5 based characters with a total 15 character length [Group A].

Group A: oooOOOQQQ111III

A random string of a radix 5 based characters with a total of 15 character length [Group B].

Group B: oooOOQQQQ11IIII

If a compression program were to be used on Group A and Group B that consisted of underlining the first individual character of a similar group of sequential characters, moving towards the right, on the string and multiplying it by a formalized system of arithmetic as found in a key, see Key Code A and Key Code B, with the compression of Group A and Group B as the final result.

Key Code A (for Group A)

$\underline{o} = \times 3$
$\underline{O} = \times 3$
$\underline{Q} = \times 3$
$\underline{1} = \times 3$
$\underline{I} = \times 3$

Group A: oOQ1I

Resulting in a 5 character length for Group A.

Key Code B (for Group B)

$\underline{o} = \times 3$
$\underline{O} = \times 2$
$\underline{Q} = \times 4$
$\underline{1} = \times 2$
$\underline{I} = \times 4$

Group B: oOQ1I

Resulting in a 5 character length for Group B.

Both Group A (non-random) and Group B (random) have the same compression value, each group resulted in a compression value of 1/3 the total pre-compression, original, state. This contrasts traditional notions of random and non-random strings [5]. These findings are similar to Tice (2003) and have applications to both Algorithmic Information Theory and Information Theory [6].

Some other examples using Example A radix 5 characters [oOQ1I] to test random and non-random sequential strings.

The following is a non-random string of radix 5 based characters with a total of 15 character length [Group A].

Group A: oooOOOQQQ111III

A random string of a radix 5 based characters with a total of 15 character length [Group C].

Group C: oooooOQQQQQ1I

If a compression program were to be used on Group A and Group C that consists of underlining the first individual character of a similar group of sequential characters, moving towards the right, on the string and multiplying it by a formalized system of arithmetic as found in a key, see Key Code A and Key Code C, with the compression of Group A and Group C as the final result.

Key Code A (for Group A)

o = ×3
O = ×3
Q = ×3
1 = ×3
I = ×3

Group A: oOQ1I

Resulting in a 5 character length for Group A.

Key Code C (for Group C)

o = ×3

O $= \times 1$
Q $= \times 5$
1 $= \times 1$
I $= \times 1$

Group C: oOQ1I

Resulting in a 5 character length for Group C.
   This example has Group A as a non-random string and Group D as a random string using radix 5 characters for a total 15 character length.
   A non-random string of radix 5 characters with a 15 character length (Group A).

Group A: oooOOOQQQ111III

A random string of a radix 5 based characters with a total of 15 character length (Group D).

Group D: oOOOOQQ1111IIII

If a compression program were to be used on Group A and Group D that consistent of underlining the first individual character of a similar group of sequential characters, moving towards the right, on the string and multiplying it by a formalized system of arithmetic as found in a key, see Key Code A and Key Code D, with the compression of Group A and Group D as the final result.

Key Code A (for Group A)

o $= \times 3$
O $= \times 3$
Q $= \times 3$
1 $= \times 3$
I $= \times 3$

Group A: oOQ1I

Resulting in a 5 character length for Group A.

Key Code D (for Group D)

o = ×1
O = ×4
Q = ×1
1 = ×4
I = ×4

Group D: oOQ1I

Resulting in a 5 character length for Group D.

As a final example Group A is a non-random sequential string and Group E is a random sequential string using a radix 5 characters for a total of 15 character length.

A non-random string of radix 5 based characters with a 15 character length [Group A].

Group A: oooOOOQQQ111III

A random string of a radix 5 based characters with a total of 15 character length [Group E].

Group E: ooOOOOQQQQ111II

If a compression program were to be used on Group A and Group E that consistent of underlining the first individual character of a similar group of sequential characters, moving to the right, on the string and multiplying it by a formalized system of arithmetic as found in a key, see Key Code A and Key Code E, with the compression of Group A and Group E as the final result.

Key Code A (for Group A)

o = ×3
O = ×3
Q = ×3
1 = ×3
I = ×3

Group A: oOQ1I

Resulting in a 5 character length for Group A.

Key Code E (for Group E)

o = ×2
O = ×4
Q = ×4
1 = ×3
I = ×2

Group E: oOQ1I

Resulting in a 5 character length for Group E.

Again, these examples conflict with traditional notions of random and non-random sequential strings in that the compression ratio is one third that of the original character number length for both the random and non-random sequential strings using a radix 5 based system.

Part III

Traditional information based systems use a binary based system represented by either a 1 or a 0. First developed by Claude Shannon in 1948 and termed 'information theory', this fundamental unit has become the backbone of our information age. One important aspect to information theory is that of data compression, the removal of redundant features in a message that can reduce the overall size of a message [7]. With the substantial compression values found in using a radix 5 based system it seems a new paradigm has arrived to carry the future of information.

Information technology has been the major drive of the economic growth in the past decade adding $2 trillion a year to the economy [8]. This growth needs to be sustained in order for new jobs and the economy to maintain a high standard of living. Only by considering alternative developments to existing models of technology, can the future of the economy develop and continue at a successful level of growth.

The internet was an outgrowth of the cold war as a government-sponsored project to develop communications network that was decentralized [9]. Today the internet is the major highway of global information with search engine technology rapidly taking center stage on both universities research departments as well as the Dow Jones index. The need to handle this vast and ever growing amount of information will need a fundamental change to the very nature of the structure of our information systems. It is clear that any new developments to deal with more and more information must begin at the fundamental level.

With a radix 5 base that has been proven to have the compression ratio similar in both random and non-random states, the question of usage as a medium for transmission and storage of information becomes paramound. With an ever increasing need for transmission and storage in the areas of telecommunications and computer science, the viability of a new system at the fundamental level of communication theory that is both robust and diverse enough to allow for future growth beyond the binary based system in use today.

Summary

This paper has shown that a radix 5 based system has profound properties of compression that are well beyond those found in binary systems using sequential strings of a random and non-random types. These compression values have strong potential applications to information theory and communication theory as a whole.

When comparing the radix 2 and the radix 5 based systems the greater compression factor of the radix 5 based system has strong applications to signal transmission and storage issues.

While the identical compression values for random and non-random radix 5 based strings is a result of this paper, the application of this theory to communication theory cannot be understated. It has been shown that a radix based 5 based system has a compression factor that makes it an ideal functional standard for future information systems, particularly in the fields of telecommunications and computer science.

References

[1] G. Ifrah, *The Universal History of Numbers*, John Wiley & Sons, New York, 2000, p. 47.
[2] H. Weyl, Symmetry, in J.R. Newman (Ed.), *The World of Mathematics*, Simon and Schuster, New York, 1956, pp. 671–724 [p. 710].
[3] C.E. Shannon, A mathematical theory of communication, *Bell Sys. Tech. J.* **27**, 1948,

379–423 & 623–656.

[4] R.K. Richards, *Arithmetic Operations in Digital Computers*, D. Van Nostrand Company, Princeton, 1955, p. 184.

[5] S. Kotz and N.I. Johnson, *Encyclopedia of Statistical Science*, John Wiley & Sons, New York, 1982, p. 39.

[6] B.S. Tice, *Two Models of Information*, 1st Book Publishers, Bloominton, 2003.

[7] B. Gates, N. Myhrvold and P. Rinearson, *The Road Ahead*, Viking, New York, 1995, p. 30.

[8] F. Davis, Impact of information technology routed, *Silicon Valley.com*, March 14, 2007, p. 1.

[9] J.E. Nuecherlein and P.J. Weiser, *Digital Crossroads*, The MIT Press, Cambridge, 2005, p. 129.

Appendix F
The Analysis of Binary, Ternary and Quaternary Based Systems for Communications Theory*

Abstract

The implementation of a ternary or quaternary based system to information infrastructure to replace the archaic binary system. Using a ternary or a quaternary based system will add greater robustness, compression, and utilizability to future information systems.

Keywords: Radix 2, Radix 3, Radix 4, Binary, Ternary, Quaternary, Information Theory, Communication Theory.

Introduction

With the advent of the superior compression of both a ternary and quaternary based system over that of the traditional binary system in information theory, the real need for a practical application to the fundamental structure of 'information' must be re-considered for the 21st century. With information technology being the 'major driver' of economic growth in the past decade, adding $2 trillion a year to the economy, the need to sustain and increase economic growth becomes an imperative [1].

With a growing interest in 'rebuilding' the internet, the fundamental question arises 'why be tied to an archaic binary based system when both a ternary and a quaternary based system are more robust, offer greater utilizability, and have far greater capacity for compression?' [2–4]. The answer to this question

---

*A poster presented at the SPIE Symposium on Optical Engineering and Applications, held in San Diego, California, August 10–14, 2008.

lies with the political aspect of the innovation process. If such a system is to be built using a ternary or a quaternary based system over the out-dated binary based system, then the government must be informed of the value of such systems over the existing system of information based infrastructures [5].

Part I

Information based systems use a binary based system represented by either a 1 or a 0. First developed by Claude Shannon in 1948 and termed 'information theory', this fundamental unit has become the 'backbone' of our information age [6]. One important aspect to information theory is that data compression, the removal of redundant features in a message, can 'reduce' the overall size of a message [6]. The need for better compression of messages is an ever growing necessity in both computing and communications [6]. The 2007 Nobel Prize for Physics was awarded for the discovery of GMR that has increased the capacity of computer hard drives [7]. A more profound effect to the computer industry would be the change from a binary based system to a ternary or quaternary based system.

The internet was a by-product of the 'cold war'. A government sponsored project to develop a communications system that was decentralized [8]. Under the Department of Defense's Advanced Research Project Agency (DARPA), the internet started life as ARPAnet in 1969 [8]. Even Tim Berners-Lee, the 'father of the web', states that "the Web is far from 'done'" and that it is a "jumbled state of construction" [9].

The out-growth of a Ph.D. dissertation, Google, the search engine company, with perhaps the most extensive computing platform in existence, wants to become an information giant [10]. Google is in some respect a 'Money Machine' with a value of $23 billion when it first hit the stock market in 2004 and has recorded an annual profit of $3 billion in 2006 [11, 12].

Part II

The advantages of a ternary and quaternary based system over a binary based system for information theory.

Radix 2 Base

Radix 2 Based System

Group A
Binary non-random sequence

[111000111000111]

Group B
Binary random sequence

[111001100011111]

If Group A and Group B are compressed using the first character type and the following similar character types in a sequential order that follows that first character type, a numerical value to the number of character types can be assigned from that similar sequence of characters that can be represented by a multiple of that number represented in that group. An example will be that [111] equals the character type 1 multiplied by three to equal [111]. Notice that the character type is not a numerical one and does not have a semantic value beyond being different than the other character type [0].

Using a Key Code as an index of which character is to be multiplied, and by what amount, a compressed version of the original length of characters results.

Key Code A (Group A)

1 = ×3
0 = ×3

Group A
Binary non-random sequence

[10101]

Resulting in Group A having a compression one third the original character length of 15 characters.

Key Code B (Group B)

1 = ×3
0 = ×3

Group B
Binary random sequence

[1_00110_11111]

Resulting in Group B having a compression two thirds the original character length of 15 characters.

A Radix 3 Base

Radix 3 Based System

If a ternary system, or radix 3 based system, was used to represent both random and non-random sequential strings, the following three character symbols can be used: [1], [0] and [Q].

Group A
A non-random ternary sequence

[111000QQQ111000QQQ]

Total character length of 18 characters.

Group C
A random ternary sequence

[111000QQQQ11000QQQQ]

Total character length of 18 characters.

Again use of a Key Code to compress the original total character length by use of multiplication.

Key Code A (Group A)

$1 = \times 3$
$0 = \times 3$
$Q = \times 3$

Group A
Non-random ternary sequence

[10Q10Q]

Total compression for Group A is a length of 6 characters from the original 18 character length. This is one third the original character length.

Key Code C (Group C)

1 = ×3
0 = ×3
Q = ×4

Group C
Random ternary sequence

[1_0_Q_110_Q_]

Total compression for Group C is a length of 7 characters from the original 18 characters length. This is less than one half of the original character length.

Radix 4 Base

Radix 4 Based System

If a quaternary, or radix 4 based system, was used to represent both random and non-random sequential strings, the following character symbols can be used: [1], [0], [Q], and [I].

Group A
A non-random quaternary sequence

[111000QQQIII111000QQQIII]

A total character length of 24 characters.

Group D
A random quaternary sequence

[1110000QQQIII111100QQIII]

Total character length of 24 characters.

The use of a Key Code to compress the original character length by use of multiplication.

Key Code A (Group A)

1 = ×3 0 = ×3 Q = ×3 I = ×3

Groun A
A non-random quaternary sequence

[10QI10QI]

Total compression for Group A is a length of 8 characters from the original 24 character length. This is one third the original character length.

Key Code D (Group D)

1 = ×
0 = ×
Q = ×
I = ×

Groun A
A non-random quaternary sequence

[1110_Q_I_1_00QQI]

Total compression for Group D is a length of 12 characters from the original 24 character length. This is one half the original character length.

Part III

In 2000 the 'Milenium Bug', or Y2K problem, arose from the perceived problem of information systems changing from one century mark to another. The concern over this problem was global in scope. Imagine the entire information system of the world being made 'redundant' by a superior information system? The concern I have for the United States is that a foreign power will implement a ternary or quaternary based information system that will 'outdate' existing binary based systems. The reason for this paper is to educate policy makers to the potential power of both a ternary and quaternary based information systems [5].

Summary

The results of using a compression engine to compress both random and non-random sequential strings of radix 2, radix 3 and radix 4 based strings resulted in the following:

Radix Base		Random	Non-random
Radix 2	15 character length total	11	5
Radix 3	18 character length total	7	6
Radix 4	24 character length total	12	8

Both the radix 3 and radix 4 based systems has substantial compression values in the random sequential strings categories. As random sequential strings have the most applicable nature to practical modes of information transmission and storages, these findings have both theoretical and applied aspects to communication theory in all of its manifestations.

References

[1] F. Davis, Impact of information technology touted, *Silicon Valley.com*, Wednesday, March 14, 2007, 1.

[2] A. Jesdanun, Rebuilding the internet, *The Modesto Bee*, Thursday, April 19, 2007, D-1 & D-3.

[3] B.S. Tice, A ternary based system for information theory, Technical Paper, *Advanced Human Design*, Volume 1, Number 1, December 2006, 1–3.

[4] B.S. Tice, A radix 4 based system for communications theory, Technical Paper, *Advanced Human Design*, Volume 1, Number 2, December 2006, 1–3.

[5] S. Boehlert, Explaining science to power: Make it simple, make it pay, *Science*, Volume 314, November 24, 2006, 1228–1229.

[6] B. Gates, N. Myhnvold and P. Rinearson, *The Road Ahead*, Viking, New York, 1995.

[7] A. Cho, Effect, Effect that revolutionized hard drives nets a nobel, *Science*, Volume 318, October 12, 2007, 179.

[8] J.E. Nuecherlein and P.J. Weiser, *Digital Crossroads*, The MIT Press, Cambridge, 2005.

[9] T. Berners-Lee, *Weaving the Web*, Harper, San Francisco, 1999.

[10] J. Battelle, *The Search*, Portfolio, New York, 2005.

[11] D.A. Vise and M. Malseed, *The Google Story*, Delacorte Press, New York, 2005.

[12] P. Durman, Man who took google global, *Times On Line*, Sunday, May 20, 2007, 1–4.

Appendix G
A Radix 4 Based System for
Information Theory*

Abstract

The paper will introduce the quaternary, or radix 4, based system for use as a fundamental standard beyond the traditional binary, or radix 2, based system in use today. A greater level of compression is noted in the radix 4 based system when compared to the radix 2 base as applied to a model of information theory.

Keywords: Radix 4, Quaternary, Information Theory, Communication Theory.

I. Introduction

A quaternary, or radix 4 based, system is defined as four separate characters, or symbols, that no semantic meaning apart from not representing the other characters. This is the same notion Shannon gave to the binary based system upon its publication in 1948 [1]. This paper will represent research that shows the radix 4 based system to have a compression value greater than the traditional radix 2 based system in use today [2].

II. Randomness

The earliest definition for randomness in a string of 1's and 0's was defined by von Mises, but it was Martin-Lof's paper of 1966 that gave a measure to

*A poster presented at the SPIE Symposium on Optical Engineering and Applications, held in San Diego, California, August 10–14, 2008.

119

randomness by the *patternlessness* of a sequence of 1's and 0's in a string that could be used to define a random binary sequence in a string [3, 4]. This is the classical measure for Kolmogorov complexity, also known as Algorithmic Information Theory, of the randomness of a sequence found in a binary string.

III. Compression Program

The compression program to be used has been termed the *Modified Symbolic Space Multiplier Program* as it simply notes the first character in a line of characters in a binary sequence of a string and subgroups them into common or like groups of similar characters, all 1's grouped with 1's and all 0's grouped with 0's, in that string and is assigned a single character notation that represents the number found in that sub-group, so that it can be reduced, compressed, and decompressed, expanded, back to its original length and form [2]. An underlined 1 or 0 is usually to note the notation symbol for the placement and character type in previous applications of this program. An underlined space following the character to be compressed will be used for this paper.

IV. Application of Theory

The application of a quaternary, or radix 4 based, system to existing communication systems has many advantages. The first is the greater amount of compression from this base, as opposed to the standard binary system in use today, and secondly, as a more utilizable system because of the four character, or symbol, based system that provides for more variety to develop information applications. From telecommunications to computing, the ternary based system applied at a fundamental standard would allow for a more robust communications system than is currently used today.

Radix 2 Base

Radix 2 Based System

Group A
Binary non-random sequence

[111000111000111]

Group B
Binary random sequence

[111001100011111]

If Group A and Group B are compressed using the first character type and the following similar character types in a sequential order that follows that first character type, a numerical value to the number of character types can be assigned from that similar sequence of characters that can be represented by a multiple of that number represented in that group. An example will be that [111] equals the character type 1 multiplied by three to equal [111]. Notice that the character type is not a numerical one and does not have a semantic value beyond being different than the other character type [0].

Using a Key Code as an index of which character is to be multiplied, and by what amount, a compressed version of the original length of characters results.

Key Code A (Group A)

1 = ×3
0 = ×3

Group A
Binary non-random sequence

[10101]

Resulting in Group A having a compression one third the original character length of 15 characters.

Key Code B (Group B)

1 = ×3
0 = ×3

Group B
Binary random sequence

[1_00110_11111]

Resulting in Group B having a compression two thirds the origihnal character length of 15 characters.

The following is an example of compression of a random and non-random radix 2 based system. A non-random sequential string of characters will have a total length of 15 characters as seen in Group A.

Group A: 111000111000111

A random sequential string of characters will have a total length of 15 characters as seen in Group B.

Group B: 110000111110111

If a compression program were to be used on Group A and Group B that consistent of underlining the first individual character of a similar group of sequential characters, moving towards the right, on the string and multiplying it by a formalized system of arithmetics as found in a key, see Key Code 1 and Key Code 2, with the compression of Group A and Group B as a final result.

Key Code 1 (for Group A)

$1 = \times 3$
$0 = \times 3$

Group A: 10101

Resulting in a compressed state of 5 characters for Group A.

Key Code 2 (for Group B)

$1 = \times 5$
$0 = \times 4$

Group B: 11<u>01</u>0111

Resulting in a compressed state of 8 characters for Group B.

Compression values of the non-random binary sequential string are one third the original 15 character length and the random binary sequential string are almost half of the original random 15 character length.

Radix 4 Base

Radix 4 based system

If a quaternary, or radix 4 based system, was used to represent both random and non-random sequential strings, the following character symbols can be used: [1], [0], [Q], and [I].

Group A
A non-random quaternary sequence

[111000QQQIII111000QQQIII]

Total character length of 24 characters.

Group D
A random quaternary sequence

[1110000QQQIII111100QQIII]

Total character length of 24 characters.

The use of a Key Code to compress the original character length by use of multiplication.

Key Code A (Group A)

1 = ×3
0 = ×3
Q = ×3
I = ×3

Group A
Non-random quaternary sequence

[10QI10QI]

Total compression for Group A is a length of 8 characters from the original 24 character length. This one third the original character length.

Key Code D (Group D)

1 = ×
0 = ×
Q = ×
I = ×

Group D
Random quaternary sequence

[1110_Q_I_1_00QQI_]

Total compression for Group D is a length of 12 characters from the original 24 character length. This is one half the original character length.

Comparison of a Radix 2 and Radix 4 Based Systems

The results of using a compression engine to compress both random and non-random sequential strings of radix 2 and radix 4 based strings resulting in the following:

Radix Base		Random	Non-random
Radix 2	15 character length total	11	5
Radix 4	24 character length total	12	8

The radix 4 based systems had substantial compression values in the random sequential strings categories. As random sequential strings have the most applicable nature to practical modes of information transmission and storage, these findings have both theoretical and applied aspects to communication theory in all of its manifestations.

Economic Issues

With the advent of the superior compression of a quaternary based system over that of the traditional binary system in information theory, the real need for a practical application to the fundamental structure of 'information' must be re-considered for the 21st century. With information technology being the 'major driver' of economic growth in the past decade the need to sustain and increase economic growth becomes an imperative.

With a growing interest in 'rebuilding' the internet, the fundamental question arises 'why be tied to an archaic binary based system when a quaternary based system is more robust, is more utilizable, and have far greater capacity for compression?' The answer to this question lies with the political aspect of the innovation process. If such a system is to be built using a quaternary based system over the outdated binary based system, then the government must be informed of the value of such systems over the existing system of information based infrastructures.

Summary

This paper has shown that a radix 4 based system has profound properties of compression that are well beyond those found in binary systems using sequential strings of a random and non-random types. These compression values have strong potential applications to information theory and communication theory as a whole.

When comparing the radix 2 and the radix 4 based systems the greater compression factor of the radix 4 based system has strong applications to signal transmission and storage issues.

It has been shown that a radix 4 based system has a compression factor that makes it an ideal functional standard for future information systems, particularly in the fields of telecommunications and computer science.

References

[1] C.E. Shannon, A mathematical theory of communication, *Bell Sys. Tech. J.* **27**, 379–423 & 623–656 (1948).
[2] B.S. Tice, A radix 4 based system for communications theory, Technical Paper, *Advanced Human Design*, Volume 1, Number 2, 1–3, December 2006.
[3] M. Li and P.M.H. Vitanyi, *An Introduction to Kolmogorov Complexity and Its Applications*. Springer, New York, 1993/1997.
[4] P. Martin-Lof, The definition of random sequences, *Information and Control* **9**, 602–619, 1966.

Bibliography

Aaronson, S. (2001) Book review 'A New Kind of Science'. *Quantum Information and Computation*, Volume 1, pp. 95–108.

ACM Press (1987) *ACM Turing Award Lectures: The First Twenty Years*. Menlo Park, California: Addison-Wesley Publishing Company. McCarthy, J., Generality in artificial intelligence, pp. 257–267.

Agar, J. (2006) Secret giants. *Nature*, Volume 442, August 17, p. 746.

Akaike, H. (1973) Information theory as an extension of the maximal likelihood principle. In B.N. Petrov and F. Csaki (Eds.) *Second International Symposium on Information Theory*, Budapest: Akademiai Kiado, pp. 267–281.

Al-Khwarizmi (820 A.D.) *The Compendious Book on Calculation by Completion and Balancing*. (See Endnote #16).

Ash, R.B. (1990) *Information Theory*. New York: Dover.

Asher, R.E. and Simpson, J.M.Y. (1994) *The Encyclopedia of Language and Linguistics: Volume I*. New York: Pergamon Press.

Atkins, P. (2003) *Galileo's Finger*. Oxford: Oxford University Press.

Baldwin, C.Y. and Clark, K.B. (2000) *Design Rules*. Cambridge, MA: The MIT Press.

Bardzin, Y.M. (1968) Complexity of programs to determine whether natural numbers not greater than n belong to a recursively enumerable set. *Soviet Mathematical Dokl.*, Volume 9, Number 5, pp. 1251–1254.

Barney, H.L. (1945) Narrow band frequency shift transmissions using 2, 4, and 8 valued signals. Internal Bell Laboratory Memorandum, July 24, 1945. Cited from Fagen, M.D. (1978), p. 316.

Baronchelli, A., Caglioti, E. and Loreto, V. (2005) Artificial sequences and complexity measures. *Journal of Statistical Mechanics: Theory and Experiment*, Issue 4, p. 04002.

Barzdin, Y.M. (1988) Algorithmic Information Theory. Entry in *Encyclopedia of Mathematics*. Dordrecht: Reidel, pp. 140–142.

Barron, I. And Curnow, R. (1979) *The Future with Microelectronics*. London: Frances Pinter.

BBC News (2006) Science 'not for normal people'. Date: January, 20th, 2006, 2 pages. Web address: http://news.bbc.couk/2/hi/uk-news/education/4630808.stm.

BBC News (2006) Last post for the telegram? Date: February 4th, 2006, 4 pages. Web address: http://news.bbc.co.uk/2/hi/americas/4674782.stm.

Bellac, M.L. (2006) *A Short Introduction to Quantum Information and Quantum Computation*. Cambridge: Cambridge University Press.

Beltrami, E. (1999) *What is Random?* New York: Copernicus.

Benedetto, D., Caglioti, E., Loreto, V. and Pietronero, L. (2002) Data compression and information in data sequences: Examples and implications for geophysical data. American Geophysical Union, Fall Meeting 2002.

Bennett, C.H., Gacs, P., Li, M., Vitanyi, P.M.B. and Zurek, W.H. (1993) Information distance. In *Proceedings of the 25th ACM Symposium on the Theory of Computing*, pp. 21–30.

Bennett, D.J. (1998) *Randomness*. Cambridge: Harvard University.

Berkeley, E.C. (1949) *Giant Brains: Or Machines That Think*. New York: John Wiley & Sons.

Berlinski, D. (2000) *The Advent of the Algorithm*. New York: Harcourt.

Berthiaume, A., van Dam, W. and Laplante, S. (2001) Quantum Kolmogorov Complexity. Proceedings of the 15th IEEE Annual Conference on Computational Complexity. *Journal of Computer and Systems Sciences*, Volume 63, Number 2, September, p. 201–221.

Biggar, H. (2006) Cupertino casts spell on computer spellcheckers. Cupertino Courier, Wednesday October 18, pp. 1–2. Webaddress:http://www.community-newspapers.com/archives/cupertinocourier/20061018/news2.shtml.

Bitter, G.G. (Editor) (1992) *MacMillan Encyclopedia of Computers: Volume 1*. New York: Macmillan Publishing Company.

Blum, L. (1967) A machine independent theory of the complexity of recursive functions. *Journal of the Association for Computing Machinery*, Volume 14, pp. 322–336.

Bol'shakov, I.A. and Smirnov, A.V. (2005) Text compression methods. *Journal of Mathematical Sciences*, Volume 56, Number 1, August, pp. 2249–2262.

Boole, G. (1847/1998) *The Mathematical Analysis of Logic*. Bristol: Thoemmes Press.

Boole, G. (1854/1951) *An Investigation of the Laws of Thought*. New York: Dover.

Brooks, F.P., Blaauwi, G.A. and Buchholz, W. (1959) Processing in bits and pieces. *IRE Transactions on Electronic Computers*, Volume EC-8, Issue 2, June, pp. 118–124.

Brillouin, L. (1956/1962) *Science and Information Theory*. New York: Academic Press.

Brooks, M. (2003) Curiouser and curiouser. *New Scientist*, May 10, pp. 28–31.

Brown, J. (2001) *The Quest for the Quantum Computer*. New York: Simon & Schuster.

Buchanan, M. (2006) What lies beneath it all. *The New Scientist*, January 28, p. 47.

Buchholz, W. (1959) Fingers or fists? *Communications of the ACM*, Volume 2, Number 12, December, pp. 3–11.

Burks, A.W., Goldstine, H.H. and von Neumann, J. (1946) Preliminary discussion of the logical design of an electronic computing instrument. In *John von Neumann Collected Works, Volume V*, A.H. Taub (Ed.). New York: The MacMillan Company, 1963, pp. 34–79.

Butterworth, B. (1999) *What Counts*. New York: The Free Press.

Calude, C.S. and Zimand, M. (2008) Algorithmically independent sequences. arXiv: 0802.0487v1, February 4.

Campbell, J. (1982) *Grammatical Man: Information, Entropy, Language and Life*. New York: Simon and Schuster.

Capra, F. (1983) *The Tao of Physics*. New York: Shambhala.

Capra, F. (1997) *The Web of Life*. New York: Anchor.

Carnap, R. (1950) *Logical Foundations of Probability*. Chicago: The University of Chicago Press.

Casti, J.L. (1996) *Five Golden Rules*. New York: John Wiley & Sons.

Casti, J. and Karlqvist, A. (2003) *Art and Complexity*. New York: Elsevier.

Cathey, S.G. (1984) Data compression for noiseless channels. Ph.D. Thesis Clemson University, South Caroline USA.

Cauley, L. (2005) *End of the Line: The Rise and Fall of AT&T*. New York: Free Press.

Chaitin, G.J. (1966) On the length of programs for computing finite binary sequences by bounded-transfer Turing machines. *AMS Notices*, Volume 13, p. 133.

Chaitin, G.J. (1966) On the length of programs for computing finite binary sequences by bounded-transfer Turing machines II. *AMS Notices*, Volume 13, pp. 228–229.

Chaitin, G.J. (1966) On the length of programs for computing finite binary sequences. *Journal of the ACM*, Volume 13, pp. 547–569.

Chaitin, G.J. (1969) On the length of programs for computing finite binary sequences: statistical considerations. *Journal of the ACM*, Volume 16, pp. 145–159.

Chaitin, G.J. (1969) On the simplicity and speed of programs for computing infinite sets of natural numbers. *Journal of the ACM*, Volume 16, pp. 407–422.

Chaitin, G.J. (1970) On the difficulty of computations. *IEEE Transactions on Information Theory*, Volume IT-16, pp. 5–9.

Chaitin, G.J. (1970) To a mathematical definition of 'life'. *ACMSICACT News*, Number 4, January, pp. 12–18.

Chaitin, G.J. (1970) Computational complexity and Gödel's incompleteness theorem. Technical Report #3/70. Departmento de Informatica, Pontificia Universidade Catolica do Rio de Janeiro, Brazil, 3/70.

Chaitin, G.J. (1970) Computational complexity and Gödel's incompleteness theorem. *AMS Notices*, Volume 17, p. 672.

Chaitin, G.J. (1971) Computational complexity and Gödel's incompleteness theorem. *ACM-SIGACT News*, Number 9, April, pp. 11–12.

Chaitin, G.J. (1972) Information-theoretic aspects of the Turing degrees. *AMS Notices*, Volume 19, pp. A-601, A-602.

Chaitin, G.J. (1972) Information-theoretic aspects of Post's construction of a simple set. *AMS Notices*, Volume 19, p. A-712.

Chaitin, G.J. (1972) On the difficulty of generating all binary strings of complexity less than n. *AMS Notices*, Volume 19, p. A-764.

Chaitin, G.J. (1973) On the greatest number of definitional or information complexity less than or equal to n. *Recursive Function Theory: Newsletter*, Number 4, January, pp. 11–13.

Chaitin, G.J. (1973) A necessary and sufficient condition for an infinite binary string to be recursive. *Recursive Function Theory: Newsletter*, Number 4, January, p. 13.

Chaitin, G.J. (1973) There are few minimal descriptions. *Recursive Function Theory: Newsletter*, Number 4, January, p. 14.

Chaitin, G.J. (1973) Information-theoretic computational complexity. *Abstracts of Papers, 1973 IEEE International Symposium on Information Theory*, p. F1-1.

Chaitin, G.J. (1974) Information-theoretic computational complexity. *IEEE Transactions on Information Theory*, Volume IT-20, pp. 10–15.

Chaitin, G.J. (1974) Information-theoretic limitations of formal systems. *Journal of the ACM*, Volume 21, pp. 403–424.

Chaitin, G.J. (1974) A theory of program size formally identical to information theory. Technical Report. International Business Machines Thomas J. Watson Research Center, Yorktown Heights, New York, pp. 1–31.

Chaitin, G.J. (1974) A theory of program size formally identical to information theory. *Abstracts of Papers. 1974 IEEE International Symposium on Information Theory*, p. 2.

Chaitin, G.J. (1975) Randomness and mathematical proof. *Scientific American*, Volume 232, Number 5, May, pp. 47–52.

Chaitin, G.J. (1975) A theory of program size formally identical to information theory. *Journal of the ACM*, Volume 22, pp. 329–340.

Chaitin, G.J. (1976) Information-theoretic characterizations of recursive infinite strings. *Theoretical Computer Science*, Volume 2, pp. 45–48.

Chaitin, G.J. (1976) Agorithmic entropy of sets. *Computers & Mathematics with Applications*, Volume 2, pp. 233–245.

Chaitin, G.J. (1976) Program size, oracles, and the jump operation. International Business Machines Corporation, Research Division. Research Report RC 5822.

Chaitin, G.J. (1976) A toy version of the LISP program. International Business Machines Corporation, Research Division. Research Report RC 5924.

Chaitin, G.J. (1977) Program size oracles, and the jump operation. *Osaka Journal of Mathematics*, Volume 14, pp. 139–149.

Chaitin, G.J. (1977) Algorithmic information theory. *IBM Journal of Research and Development*, Volume 21, pp. 350–359, 496.

Chaitin, G.J. (1977) Recent work on algorithmic information theory. *Abstracts of Papers, 1977 IEEE International Symposium on Information Theory*, p. 129.

Chaitin, G.J. (1977) Toward a mathematical definition of 'life' II. International Business Machines Corporation, Research Division. Research Report RC 6919.

Chaitin, G.J. (1978) A note on Monte Carlo primality tests and algorithmic information theory. *Communications on Pure and Applied Mathematics*, Volume 31, pp. 521–527.

Chaitin, G.J. (1979) Algorithmic information theory. International Business Machines Corporation, Research Division. Research Report RC 8019.

Chaitin, G.J. (1979) Toward a mathematical definition of 'life'. In *The Maximal Entropy Formalism*, R.D. Levine and M. Tribus. MIT Press, pp. 477–498.

Chaitin, G.J., Auslander, M.A., Chandra, A.K., Cocke, J., Hopkins, M.E. and Markstein, P.W. (1980) Register allocations via coloring. International Business Machines Corporation, Research Division. Research Report RC 8395.

Chaitin, G.J,. (1981) Register allocation & spilling via coloring. International Business Machines Corporation, Research Division. Research Report RC 9124.

Chaitin, G.J. (1982) Algorithmic Information Theory. *Encyclopedia of Statistical Sciences*, Volume 1. New York: John Wiley & Sons, pp. 38–41.

Chaitin, G.J. (1982) Godel's theorem and information. *International Journal of Theoretical Physics*, Volume 22, pp. 941–954.

Chaitin, G.J. (1985) Random reals and exponential Diophantine equations. International Business Machines Corporation. Research Report RC 11788.

Chaitin, G.J. (1986) Randomness and Godel's theorem. *Mondes en Development*, Numbers 54–55, pp. 125–128.

Chaitin, G.J. (1987) Incompleteness theorems for random reals. *Advances in Applied Mathematics*, Volume 8, pp. 119–146.

Chaitin, G.J. (1987b) *Algorithmic Information Theory*. Cambridge: Cambridge University Press.

Chaitin, G.J. (1987c) *Information, Randomness and Incompleteness: Papers on Algorithmic Information Theory*. Singapore: World Scientific Publishing Company.

Chaitin, G.J. (1987) Computing the busy beaver function. In *Open Problems in Communication and Computation*, T.M. Cover and B. Gopinath. New York: Springer-Verlag.

Chaitin, G.J. (1988) An algebraic equation for halting probability. In *The Universal Turing Machine*, R. Herken. Oxford: Oxford University Press.

Chaitin, G.J. (1988) Randomness and arithmetic. *Scientific American*, Volume 259, Number 1, July, pp. 80–85.

Chaitin, G.J. (1989) A computer gallery of mathematical physics. International Business Machines Corporation, Research Division. Research Report RC 14971.

Chaitin, G.J. (1989) Algorithmic information and evolution. International Business Machines Corporation, Research Division. Research Report RC 15240.

Chaitin, G.J. (1989) Undecidability and randomness in pure mathematics. International Business Machines Corporation. Research Report RC 15241,

Chaitin, G.J. (1991) Algorithmic information & evolution. In *Perspectives on Biological Complexity*, O.T. Solbrig and G. Nicolis. N.A.: IUBS Press, pp. 51–60.

Chaitin, G.J. (1990) A random walk in arithmetic. *New Scientist*, Volume 125, Number 1709, March 24, pp. 44–46.

Chaitin, G.J. (1991) Le hasard des numbers. *La Recherche*, Volume 22, Number 232, May, pp. 610–615.

Chaitin, G.J. (1991) Complexity and biology. *New Scientist*, Volume 132, Number 1789, October 15, p. 52.

Chaitin, G.J. (1991) Number and randomness. International Business Machines Corporation, Research Division. Research Report RC 16530.

Chaitin, G.J. (1992) LISP program-size complexity. *Applied Mathematics and Computation*, Volume 49, pp. 79–93.

Chaitin, G.J. (1992) Information-theoretic incompleteness. *Applied Mathematics and Computation*, Volume 52, pp. 83–101.

Chaitin, G.J. (1992) LISP program-size complexity II. *Applied Mathematics and Computation*, Volume 52, pp. 103–126.

Chaitin, G.J. (1992) LISP program-size complexity III. *Applied Mathematics and Computation*, Volume 52, pp. 127–139.

Chaitin, G.J. (1992) LISP program-size complexity IV. *Applied Mathematics and Computation*, Volume 52, pp. 141–147.

Chaitin, G.J. (1992) A diary on information theory. *The Mathematical Intelligencer*, Volume 14, Number 4, Fall, pp. 69–71.

Chaitin, G.J. (1992) *Information-Theoretic Incompleteness*. Singapore: World Scientific Publishing.

Chaitin, G.J. (1993) Exhibiting randomness in arithmetic using Mathematica and C. International Business Machines Corporation, Research Division. Research Report RC 18946.

Chaitin, G.J. (1993) The limits of mathematics: Course outline and software. International Business Machines Corporation, Research Division. Research Report RC 19324.

Chaitin, G.J. (1993) Randomness in arithmetic and the decline and fall of reductionism in pure mathematics. *Bulletin of the European Association for Theoretical Computer Science*, Volume 50, June, pp. 314–328.

Chaitin, G.J. (1993) On the number of n-bit strings with maximum complexity. *Applied Mathematics and Computation*, Volume 59, pp. 97–100.

Chaitin, G.J. (1994) Response to 'Theoretical mathematics'. *Bulletin of the American Mathematical Society*, Volume 30, pp. 181–182.

Chaitin, G.J. (1995) Program size complexity computes the halting problem. *Bulletin of the European Association for Theoretical Computer Science*, Volume 57, October, p. 198.

Chaitin, G.J. (1995) The Berry paradox. *Complexity*, Volume 1, Number 1, pp. 26–30.

Chaitin, G.J. (1995) A new version of algorithmic information theory. International Business Machines Corporation, Research Division. Research Report RC 20092.

Chaitin, G.J. (1995) How to run algorithmic information theory on a computer. International Business Machines Corporation, Research Division. Research Report RC 20193.

Chaitin, G.J. (1995/1996) A new version of algorithmic information theory. *Complexity*, Volume 1, Number 1, pp. 55–59.

Chaitin, G.J. (1996) How to run algorithmic information theory on a computer. *Complexity*, Volume 2, Number 1, September, pp. 15–21.

Chaitin, G.J. (1996) The limits of mathematics. *Journal of Universal Computer Science*, Volume 2, pp. 270–305.

Chaitin, G.J. (1997) An invitation to algorithmic information theory. In *DMTCS'96 Proceedings*. New York: Springer-Verlag, pp. 1–23.

Chaitin, G.J. (1998) *The Limits of Mathematics*. New York: Springer-Verlag.

Chaitin, G.J. (1999) Elegant LISP program. In *People and Ideas in Theoretical Computer Science*, C. Calude. New York: Springer-Verlag, pp. 32–52.

Chaitin, G.J. (1999) *The Unknowable*. Singapore: Springer-Verlag.

Chaitin, G. and Calude, C. (1999) Randomness everywhere. *Nature*, Volume 400, pp. 319–320.

Chaitin, G.J. (2000) A century of controversy over the foundations of mathematics. In *Finite vs. Infinite*, C. Calude and G. Paun. New York: Springer-Verlag, pp. 75–100.

Chaitin, G.J. (2000) A century of controversy over the foundations of mathematics. *Complexity*, Volume 5, Number 5, May/June, pp. 12–21.

Chaitin, G.J. (2001) *Exploring Randomness*. London: Springer-Verlag.

Chaitin, G.J. (2002) *Conversations with a Mathematician*. New York: Springer-Verlag.

Chaitin, G.J. (2002) Meta-mathematics and the foundations of mathematics. *Bulletin of the European Association for Theoretical Computer Science*, Volume 77, June, pp. 167–179.

Chaitin, G.J. (2002) Paradoxes of randomness. *Complexity*, Volume 7, Number 5, May/June, pp. 14–21.

Chaitin, G.J. (2002) Complexite, logique et hasard. In *La Complexite*, R. Benkirane. Editions Le Pommier, pp. 283–310.

Chaitin, G.J. (2002) The unknowable. In *Bridge the Gap?*, A. Miyake and H.U. Obrist. Dresden, Germany: Walter Konig, pp. 39–47.

Chaitin, G.J. (2003) Two philosophical applications of algorithmic information theory. In *Discrete Mathematics and Theoretical Computer Science*, C.S. Calude, M.J. Dinneen and V. Vajnovszki. New York: Springer-Verlag, pp. 1–10.

Chaitin, G.J. (2003) *From Philosophy to Program Size*. Estonia: Tallinn Institute of Cybernetics.

Chaitin, G.J. (2003) L'univers est-il intelligible? *Le Recherche*, Number 370, Decembre, pp. 34–41.

Chaitin, G.J. (2004) Thoughts on the Riemann hypothesis. *The Mathematical Intelligencer*, Volume 26, Number 1, Winter, pp. 4–7.

Chaitin, G.J. (2004) On the intelligibility of the universe and the notions of simplicity, complexity and irreducibility. *Grenzen and Grensuberschreitungen, XIX*, Deutscher Kongress fur Philosophie, Bonn, 23–27 September 2002, Vortrage and Kolloquien, Herausgegeben von Wolfram Hogrebe in Verbindung mit Joacham Bromand Akademie Verlag, Berlin, pp. 517–534.

Chaitin, G.J. (2004) Leibniz, randomness and the halting probability. *Mathematics Today*, Volume 40, pp. 138–139.

Chaitin, G.J. (2005) *Meta Math!* New York: Pantheon.

Chaitin, G.J. (2005) Algorithmic irreducibility in a cellular automata universe. *Journal of Universal Computer Science*, Volume 11, pp. 1901–1903.

Chaitin, G.J. (2006) The limits of reason. *Scientific American*, Volume 294, Number 3, March, pp. 74–81.

Chen, X, Kwong, S, and Li, M. (1999) A compression algorithm for DNA sequences and its applications in genome comparison. *Genome Information Service Workshop on Genome Information*, Volume 10, pp. 51–61.

Cherri, A.K. (1996) High-radix signed-digit arithmetic using symbolic substitution computations. *Optics Communications*, Volume 128, Issues 1–3, pp. 108–122.

Cho, D. (2006) March of the qubits. *New Scientist*, March 25, pp. 42–45.

Chomsky, A. N. (1956) Three models of the description of language. *IRE Transactions on Information Theory*, Volume IT-2, Number 3, September, pp. 113–124.

Chomsky, N. (1957) *Syntactic Structures*. The Hague: Mouton & Co.

Chomsky, N. (1965) *Aspects of the Theory of Syntax*. Cambridge: The MIT Press.

Church, A. (1940) On the concept of a random sequence. *Bulletin of the American Mathematical Society*, Volume 46, Number 12, Part 2, December, pp. 130–135.

Coles, P. (2006) *From Cosmos to Chaos: The Science of Unpredictability*. Oxford: Oxford University Press.

Conway, F. and Siegelman, J. (2004) *Dark Hero of the Information Age: In Search of Norbert Wiener The Father of Cybernetics*. New York: Basic Books.

Cook, S. (1971) The complexity of theorem proving procedures. In *Proceedings of the Third Annual ACM STOC Symposium*, pp. 151–158.

Cooper, D. and Lynch, M.F. (1981) Text compression using variable to fixed length encodings. *Journal of the American Society for Information Science*, Volume 33, Number 1, pp. 18–31.

Copeland, B.J. (2006) *Colossus: The Secrets of Bletchly Park's Codebreaking Computers*. Oxford: Oxford University Press.

Cover, T.M. (1991) *Elements of Information Theory*. New York: Wiley-Interscience.

Crutchfield, J.P. (2003) What lies between order and chaos?. In *Art and Complexity*, J. Casti and A. Karlqvist (Eds.). New York: Elsevier, pp. 31–45.

Dahl, O.J., Djkstra, E.W. and Hoare, C.A.R. (1972) *Structured Programming*. New York: Academic Press.

Dai, W. and Milenkovic, O. (2008) Subspace pursuit of compressive string: Closing the gap between performance and complexity. arXiv: 0803.0811v1, March 6.

Davisson, L.D. and Gray, R.M. (1975) Advances in data compression. *Advances in Communication Systems*, Volume 4, pp. 199–228.

Davisson, L.E. and Gray, R.M. (1976) *Data Compression*. Pennsylvania: Dowden, Hutchinson & Ross.

Deco, G. and Obradovic, D. (1996) *An Information-Theoretic Approach to Neural Computing*. New York: Springer-Verlag.

De Leeuw, K., Moore, E.F., Shannon, C.E. and Shapiro, N. (1956) Computability by probabilistic machines. In *Automata Studies*, C.E. Shannon and J. McCarthy. Princeton: Princeton University Press.

Deutsch, D. (1997) *The Fabric of Reality*. New York: The Penguin Group.

Dijkstra, E.W. (1982) *Selected Writings on Computing: A Personal Perspective*. New York: Springer-Verlag.

Ding, C., Kohel, D.R. and Ling, S. (2000) Secret-sharing with a class of ternary codes. *Theoretical Computer Science*, Volume 246, Issues 1–2, September 6, pp. 285–298.

Donoho, D.L. (2002) The Kolmogorov sampler. Department of Statistics Inprint, Stanford University, Stanford, California USA. Number 2002-4, January.

Doob, J.L. (1949) A mathematical theory of communication – C.E. Shannon. *Mathematical Reviews*, Volume 10, Number 2, p. 133.

Doob, J.L. (1959) *Mathematical Reviews*, Volume 5, Number 3, p.

Doob, J.L. (1953) *Stochastic Processes*. New York: John Wiley & Sons.

Dowling, J.P. (2006) To compute or not to compute? *Nature*, Volume 439, February 23, p. 919.

Downey, R.G. and Hirschfeldt, D. (2007) *Algorithmic Randomness and Complexity*. New York: Springer.

Dreyfus, H.L. (1972) *What Computers Can't Do: A Critique of Artificial Reason*. New York: Harper & Row.

Du, D.Z. and Ko, K.I. (2001) *Problem Solving in Automata, Languages, and Complexity*. New York: John Wiley & Sons.

Eijck, van J. (1994) Finite state machines. In *The Encyclopedia of Language and Linguistics: Volume 3*, R.E. Asher and J.M.Y. Simpson. Oxford: Pergamon Press, p. 1244.

Ernst, D. (2006) The offshoring of innovation. *Far Eastern Economic Review*, Volume 169, Number 4, May, pp. 29–33.

Fagen, M.D. (Ed.) (1975) *A History of Engineering and Science in the Bell System: The Early Years (1875–1925)*. Murray Hill: Bell Telephone Laboratories.

Fagen, M.D. (Ed.) (1978) *A History of Engineering and Science in the Bell System: National Service in War and Peace (1925–1975)*. Murray Hill: Bell Telephone Laboratories.

Falcioni, M., Loreto, V. and Vulpiani, A. (2003) Kolmogorov's legacy about entropy, chaos, and complexity. In *The Kolmogorov Legacy in Physics*, R. Livi and A. Vulpiani. New York: Springer-Verlag, pp. 103–108.

Fano, R.M. (1949) The Research Laboratory of Electronics. Technical Report Number 65, M.I.T., March 17, 1949. In Shannon, C.E. (1949/1998) *A Mathematical Theory of Communication*, Printed for the 50th Anniversary Edition for the 1998 IEEE International Symposium on Information Theory, MIT, Cambridge, MA, August 16–21, 1998, p. 21, footnote.

Fano, R.M. (1961) *Transmission of Information*. New York: The MIT Press and John Wiley & Sons.

Ferragina, P., Nitto, I. and Venturini, R. (2008) Bit-optimal Lempel–Ziv compression. arXiv:0802.0835v1, February 6.

Fine, T.L. (1973) *Theories of Probability: An Examination of Foundations*. New York: Academic Press.

Flamm, K. (1988) *Creating the Computer*. Washington, DC: The Brookings Institute.

Floyd, R.W. and Beigel, R. (1994) *The Language of Machines*. New York: Computer Science Press.

Franzen, T. (2005) *Godel's Theorem: An Incomplete Guide to Its Use and Abuse*. Wellesley, MA: A.K. Peters.

Fredkin, E. and Toffoli, T. (1982) Conservation logic. *International Journal of Theoretical Physics*, Volume 21.

Friedman, W.F. (1928) Report of the history of the use of codes and code language. *International Radio Telephone Conference of Washington: 1927*. Washington, DC: U.S. Government Printing Office.

Friedman, W.F. and Mendelsohn, C,J, (1932) Notes on code words. *American Mathematical Monographs*, Volume 39, August/September, pp. 394–409.

Gacs, P. (1974) On the symmetry of algorithmic information. *Soviet Math. Dokl.*, Volume 15, pp. 1477–1480. Note: Corrections, *Soviet Math. Dokl.*, Volume 15, Number 6.

Gacs, P. (1989) Review of Gregory J. Chaitin, Algorithmic Information Theory. *Journal of Symbolic Logic*, Volume 54, pp. 624–627.

Gacs, P. (2005) Quantum algorithmic entropy. A Paper for the Centrum voor Wiskunde en Informatica, Amsterdam, the Netherlands, pp. 1–20.

Gagie, T. (2006) Large alphabets and incompressibility. *Information Processing Letters*, Volume 99, pp. 246–251.

Gallager, R.G. (1968) *Information Theory and Reliable Communication*. New York: John Wiley & Sons.

Ge, M. (2005) Information Theory? In *The New Encyclopedia Britannica*, Volume 21. Chicago: Encyclopedia Britannica, pp. 631–637.

Gell-Mann, M. (1994) *The Quark and the Jaguar*. New York: W.H. Freeman and Company.

Gilbert, E.N. (1966) Information theory after 18 years. *Science*, Volume 152, April 15, pp. 320–326.

Gillie, A.C. (1965) *Binary Arithmetic and Boolean Algebra*. New York: McGraw-Hill Book Company.

Gillispie, C.C. (1971) *Dictionary of Scientific Biography: Volume IV*. New York: Charles Scribner's & Sons.

Glaser, A. (1971) *History of Binary and Other Nondecimal Numeration*. Los Angeles: Tomash Publishers.

Gleick, J. (1987) *Chaos: Making a New Science*. New York: Penguin Books.

Goldman, S. (2005) *Information Theory*. New York: Dover.

Goldstein, R. (2005) *Incompleteness: The Proof and Paradox of Kurt Godel*. New York: W.W. Norton & Company.

Goldstine, H.H. (1972) *The Computer from Pascal to von Neumann*. Princeton: Princeton University Press.

Goldstine, H.H. and von Neumann, J. (1946) On the principles of large scale computing machines. In *John von Neumann Collected Works, Volume V*, A.H. Taub (Ed.). New York: The Macmillan Company, 1963, pp. 1–32.

Grattan-Guiness, I. (2006) Incomplete mathematics. *Nature*, Volume 439, February 16, pp. 790–791.

Gray, L. (2003) A mathematician looks at Wolfram's new kind of science. *Notices of the AMS*, Volume 50, Number 2, February, pp. 200–211.

Grunwald, P.D., Myung, I.J. and Pitt, M.A. (2005) *Advances in Minimum Description Length.* Cambridge: The MIT Press.

Gutknecht, M.H. (1990) The pioneer days of scientific computing in Switzerland. In *A History of Scientific Computing*, S.G. Nash. Menlo Park, CA: Addison-Wesley Publishing Company.

Hamming, R.W. (1950) Error detecting and error correcting codes. *Bell System Technical Journal*, Volume 29, Number 2, April, p. 147.

Hankerson, D., Harris, G.A. and Johnson, P.D. (1998) *Introduction to Information Theory and Data Compression.* London: CRC Press.

Harris, R.A. (1993) *The Linguistic Wars.* New York: Oxford University Press.

Hartley, R.V.L. (1928) Transmission of information. *The Bell System Technical Journal*, Volume 7, Number 3, pp. 535–563.

Hartmanis, J. (1983) Generalized Kolomogorov complexity and the structure of feasible computations. In *Proceedings of the 24th IEEE Symposium on Foundations of Computer Science*, pp. 436–445.

Hayden, P., Jozsa, R. and Winter, A. (2002) Trading quantum for classical resources in quantum data compression. *Journal of Mathematical Physics*, Volume 43, Issue 9, pp. 4404–4444.

Henck, F.W. and Strassburg, B.(1988) A *Slippery Slope: The Long Road to the Breakup of AT&T.* West Port, CT: Greenwood Press.

Hey, A. (1999) *Feynman and Computation.* Reading, MA: Perseus Books.

Hey, A. and Allen, R.W. (1999) *Feynman Lectures in Computation.* Nashville, TN: Westview.

Hiltzik, M. (1999) *Dealers of Lightening.* New York: Harper Business.

Hochfelder, D. (2002) Constructing an industrial divide: Western Union, AT&T, and the Federal Government, 1876–1971. *Business History Review*, Volume 76, Issue 4, March 31, pp. 705–732.

Horgan, J. (1996) *The End of Science.* New York: Helix Books.

Horizon (1962) A new American poet speaks: The work of A.B. *Horizon*, Volume IV, Number 5, May, pp. 96–99.

Hromkovic, J. (2004) *Theoretical Computer Science.* New York: Springer.

Huffman, D.A. (1952) A method for the construction of minimum redundancy codes. *Proceedings of the IRE*, Volume 40, Number 9, September, pp. 1098–1101.

Hugill, P.J. (1999) *Global Communications Since 1844.* Baltimore: The John Hopkins University Press.

Hurford, J.R. (1975) *The Linguistic Theory of Numerals.* Cambridge: Cambridge University Press.

Hutter, M. (2005) *Universal Artifical Intelligence.* New York: Springer.

Ifrah, G. (2000) *The Universal History of Numbers.* New York: John Wiley & Sons.

Ifrah, G. (2001) *The Universal History of Computing.* New York: John Wiley & Sons.

Jensen, O. (1963) Who's underdeveloped? *Horizon*, Volume V, Number 3, January, pp. 116–117.

Johnson, G. (2000) *Strange Beauty: Murrey Gell-Mann and the Revolution in Twentieth-Century Physics.* New York: Alfred A. Knopf.

Johnson, G. (2003) *A Shortcut Through Time.* New York: Alfred A. Knopf.

Jones, N.D. (1997) *Computability and Complexity.* Cambridge: The MIT Press.

Jones, G.A. and Jones, J.M. (2000) *Information and Coding Theory.* London: Springer-Verlag.

Kahre, J. (2002) *The Mathematical Theory of Information.* New York: Springer.

Kalnishkan, Y., Vovk, V. and Vyugin, M.V. (2005) How many strings are easy to predict? *Information and Computation,* Volume 201, Issue 1, August 25, pp. 55–71.

Kanovich, M.I. (1970) Complexity of resolution of a recursively enumerable set as a criterion of its universality. *Soviet Math. Dokl.,* Volume 11, Number 5, pp. 1224–1228.

Karp, R. and Lipton, R. (1982) Turing machines that take advise. *Ensiegn. Math.,* Volume 28, pp. 192–209.

Kaye, A.S. (1990) Arabic. In *The World's Major Languages,* B. Comrie (Ed.). New York: Oxford University Press, Chapter 33, pp. 664–685.

Kemp, M. (2006) Intimations and intuitions. *New Scientist,* September, pp. 48–49.

Khinchin, A.I. (1957) *Mathematical Foundations of Information Theory.* New York: Dover Publications.

Klir, G.J. (2006) *Uncertainty and Information.* New Jersey: John Wiley & Sons.

Knuth, D.E. (1960) n/a. *CACM,* Volume 3, pp. 245–247.

Knuth, D.E. (1998) *The Art of Computer Science. Volume 2: Semi-Numerical Algorithms.* Menlo Park, CA: Addison-Wesley.

Kolomogorov, A.N. (1933/1956) *Foundations of the Theory of Probability.* Chelsea. English translation by N. Morrison.

Kolmogorov, A.N. (1941) Interpolation and extrapolation. *Bulletin de l'Academie des Sciences de USSR, Ser. Math.,* Volume 5, pp. 3–14.

Kolmogorov, A.N. (1965) Three approaches to the quantitative definition of information. *Problems of Information Transmission,* Volume 1, Number 1, January–March, pp. 1–7.

Kolmogorov, A.N. (1968) Logical bases for information theory and probability theory. *IEEE Transactions Information Theory,* Volume IT-14, September, pp. 662–664.

Kolmogorov, A.N. (1969) On the logical foundations of information theory and probability theory. *Problems of Information Transmission,* Volume 5, Number 3, July–September, pp. 1–4.

Lambalgen, M. van (1988) Algorithmic information theory. Instituut voor Taal, Logica en Informatie, Delft University of Technology, Amsterdam, the Netherlands, pp. 1–17.

Landauer, R. (1961) Irreversibility and heat generation in the computing process. *IBM Journal of Research and Development,* Volume 5, pp. 183–191.

Landauer, R. (1993) Information is physical. In *Proceedings of the Workshop on Physics and Computation, PhysCom'92.* Los Alamitos: IEEE Computer Society Press, pp. 1–4.

Landauer, R. (1984) Fundamental physical limitations of the computational process. In *Computer Culture: The Scientific, Intellectual, and Social Impact of the Computer.* New York: The New York Academy of Sciences, pp. 161–170.

Lanier, J. (2006) Two philosophies of mathematical weirdness. *American Scientist,* May–June, pp. 269–271.

Lanouette, W. and Silard, B. (1994) *Genius in the Shadows.* Chicago: The University of Chicago Press.

Lassaigne, R. and de Rougemont, M. (2004) *Logic and Copmplexity.* New York: Springer.

Le Bellac, M. (2006) *A Short Introduction to Quantum Information and Quantum Computation.* Cambridge: Cambridge University Press. (Translated from the French by P. Forerand-Millard).

Lecuyer, C. (1999) Making Silicon Valley: Engineering, Culture, Innovation, and Industrial Growth, 1930–1970. Ph.D. Dissertation, Stanford University, 302 pp. U.M.I. Note: Expanded and published as *Making Silicon Valley: Innovation and the Growth of High Tech, 1930–1970*, 2006, MIT Press, 424 pp.

Lecuyer, C. (2006) *Making Silicon Valley: Innovation and the Growth of High Tech, 1930–1970*. Cambridge: The MIT Press.

Levin, J. (2006) The truth of lies. *New Scientist*, August 10, pp. 44–45.

Levin, L.A. (1974) laws of information conservation (no growth) and aspects of the foundation of probability theory. *Problems with Information Transmission*, Volume 10, pp. 206–210.

Levy, S. (2006) *The Perfect Thing: How the iPod Shuffles Commerce, Culture, and Coolness*. New York: Simon and Schuster.

Li, M. and Vitanyi, P.M.B. (1985) Tape verses queue and stacks: The lower bounds. Centrum voor Wiskunde en Informaticsa, Amsterdam, the Netherlands, CS-R8519.

Li, M., Longpre, L. and Vitanyi, P.M.B. (1986) The power of queue. Centrum voor Wiskunde en Informatica, Amsterdam, the Netherlands, CS-R8612.

Li, M., Longpre, L. and Vitanyi, P.M.B. (1986a) The poer of the queue. Massachusetts Institute of Technology. Laboratory for Computer Science, MIT/LCS/TM-303.

Li. M. and Vitanyi, P.M.B. (1988) Two decades of applied Kolmogorov complexity: In memoriam Andrei Nikolaevich Kolmogorov, 1903–1987. Centrum voor Wiskunde en Informatica, Amsterdam, the Netherlands, CS-R8813.

Li, M. and Vitanyi, P.M.B. (1989) Kolmogorov complexity and its applications. Centrum voor Wiskunde en Informatica, Amsterdam, the Netherlands, CS-R8901.

Li, M. and Vitanyi, P.M.B. (1989a) Inductive reasoning and Kolmogorov complexity. Centrum voor Wiskunde en Informatica, Amsterdam, the Netherlands, CS-R8915.

Li, M. and Vitanyi, P.M.B. (1989b) A new approach to formal language theory by Kolmogorov complexity. Centrum voor Wiskunde en Informatica, Amsterdam, The Netherlands, CS-R8919.

Li, M. and Vitanyi, P.M.B. (1991) Combinatorics and Kolmogorov complexity. Centrum voor Wiskunde en Informatica, Amsterdam, the Netherlands, CS-R9125.

Li, M. and Vitanyi, P.M.B. (1993/1997) *An Introduction to Kolmogorov Complexity and Its Applications*. New York: Springer.

Li, Y., Song, F., Wang, H., Jin, H and Zhang, H. (2008) A method for compressing off-axis digital hologram with spectrum shifting. *Proceedings of SPIE*, Volume 6832, p. 68320A.

Lin, S.C., Athale, R.A. and Lee, J.N. (1983) A compact optical pulse compressor with computer generated holographic masks for radar applications. Progress Report, April 23–September 30, 1982 Naval Research Laboratory, Washington, DC, Applied Optics Branch.

Liu, Z., Xiong, Z., Wu, Q., Wang, Y. and Castleman, K.R. (2001) Human chromosome image compression using cascaded differential and wavelet coding. In *Proceeding of SPIE Applications of Digital Image Processing XXIV*, A.G. Tescher (Ed.), Volume 4472, pp. 349–360.

Lloyd, S. (2006) *Programming the Universe: A Quantum Computer Scientist Takes on the Cosmos*. New York: Knopf.

Longpre, L. (1992) Resource bounded kolmogorov complexity and statistical tests. In *Kolmogorov Complexity and Computational Complexity*, O. Watanabe (Ed.). Berlin: Springer-Verlag, pp. 66–84.

Loreto, V. and Puglisi, A. (2003) Data compression approach to sequence analysis. *Modeling of Complex Systems*, Volume 661, April 7, pp. 184–187.

Los Altos News (1965) Tinker Toys to Riches? *Los Altos News*, Volume 27, Number 12, Wednesday October 13, Front page.

Luenberger, D.G. (2006) *Information Science*. Princeton: Princeton University Press.

Lutz, J.H. (2003) The dimensions of individual strings and sequences. *Information and Computation*, Volume 187, Issue 1, November 25, pp. 49–79.

Machlup, F. and Mansfield, U. (1983) *The Study of Information: Interdisciplinary Messages*. New York: John Wiley and Sons.

Machta, J. (1999) Entropy, information and computation. *American Journal of Physics*, Volume 67, Issue 12, December, pp. 1074–1077.

MacKay, D.M. (1969) *Information, Mechanism, and Meaning*. Cambridge: MIT Press.

MacKay, D.J.C. (2003) *Information Theory, Inference, and Learning Algorithms*. Cambridge: Cambridge University Press.

Malescio, G. (2006) Coping with uncertainty. *Nature*, Volume 443, October 26, p. 918.

Malyutov, M.B. (2005) Authorship attribution of texts: A review. *General Theory of Information Transfer and Combinatorics*, Volume 21, August 1, pp. 353–357.

Markowsky, G. (1997) An introduction to algorithmic information theory. *Complexity*, Volume 2, Number 4, pp. 14–22.

Martin-Lof, P. (1966) The definition of random sequences. *Information and Control*, Volume 9, pp. 602–619.

Masani, P.R. (1989) Norbert Wiener: A survey of a fragment of his life and work. In *A Century of Mathematics in America*, R.A. Askey, H.M. Edwards and U.C. Merzbach (Eds.). Providence, RI: American Mathematical Society, pp. 299–341.

McCarthy, J. (1962) *The LISP 1.5 Programmer's Manual*. Cambridge, MA: The MIT Press.

Menninger, K. (1969) *Number Words and Number Symbols*. Cambridge, MA: The MIT Press.

Metropolis, N., Howlett, J. and Rota, G.C. (1980) *A History of Computing in the Twentieth Century*. New York: Academic Press. Programming in America in the 1950s: Some Personal Impressions, by John Backus, pp. 125–135.

Meyer, D.A. (2002) Quantum games and quantum algorithms. In *Quantum Computation and Information*, S.J. Lomonaco (Ed.), Contemporary Mathematics Series, Volume 305. Providence, Ri: American Mathematical Society, pp. 213–220.

McMurray, E.J. (1995) *Notable Twentieth-Century Scientists: Volume 2*. London: Gale Research.

Milburn, G.J. (1998) *The Feynman Processor*. Reading, MA: Perseus Books.

Millman, S. (Ed.) (1983) *A History of Science & Engineering in the Bell System: Physical Sciences, 1925–1980*. Murray Hill: AT&T Bell Laboratories.

Millman, S. (Ed.) (1984) *A History of Science & Engineering in the Bell System: Communications Sciences, 1925–1980*. Indianapolis, IN: AT&T Customer Information Center.

Milosavljevic, A. and Jurka, J. (1993) Discovering simple DNA sequences by algorithmic significance method. *Computer Application Bioscience*, Volume 9, pp. 407–411.

Mindell. D.A. (2002) *Between Human and Machine*. Baltimore: The John Hopkins University Press.

Minsky, M.L. (1967) *Computation: Finite and Infinite Machines*. Englewood Cliffs, NJ: Prentice-Hall.

Nagel, E. and Newman, J.R. (2002) *Godel's Proof*. New York: New York University Press.

Nash, S.G. (1990) *A History of Scientific Computing*. New York: ACM Press and Menlo Park, CA: Addison-Wesley Publishing Company.

Needham, R.M. (1996) Computers and communications. In *Computing Tomorrow*, I. Ward and R. Millner (Eds.). Cambridge: Cambridge University Press, pp. 284–292.

New Scientist (2001) The omega man. *New Scientist*, March 10.

New Scientist (2006) Quantum computers? Don't hold your breath. *New Scientist*, December 2, p. 17.

Oexle, K. (1995) Data compression, physical entropy, and evolutionary a priori relation between observer and object. *Physical Review*, Volume E51, Number 002651, 3 pp.

Oldfield, H.R. (1996) *King of the Seven Dwarfs: General Electric's Ambiguous Challenge to the Computer Industry*. Los Alamitos, California: IEEE Computer Society Press.

Pappou, T. and Tsangaris, S. (1997) Development of an artificial compressibility methodology using flux vector splitting. *International Journal of Numerical Methods in Fluids*, Volume 25, Issue 5, pp. 523–545.

Pavicic. M. (2003) *Quantum Computation and Quantum Communication: Theory and Experiments*. New York: Springer-Verlag.

Pierce, J.R. (1961/1980) *An Introduction to Information Theory*. New York: Dover.

Popel, D.V. (2000) Informational theoretic approach to logic functions minimization. Ph.D. Dissertation, Computer Science, Technical University of Szczecin, Poland, 161 pp.

Popular Science (date unknown) Review – 'Conversations with a Mathematician': Gregory Chaitin. *Popular Science Book Review*. (http://www.popluarscience.co.uk/reviews/rev82.htm)

Powell, D.R., Dowe, D.L., Allison, L. and Dix, T.I. (1998) Discovering simple DNA sequences by compression. *Pacific Symposium on Biocomputing*, pp. 597–608.

Purser, M. (1995) *Introduction to Error-Correcting Codes*. Boston: Artech House.

Raatikainen, P. (1998) Complexity, information and incompleteness. Doctorial Dissertation in Philosophy, University of Helsinki, Helsinki, Finland.

Raatikainen, P. (1998a) On interpreting Chaitin's incompleteness. *Journal of Philosophical Logic*, Volume 27, December, pp. 569–587.

Raatikainen, P. (2000) Algorithmic information theory and undesirability. *Syntheses*, Volume 123, pp. 217–225.

Raatikainen, P. (2001) Exploring randomness and the unknowable. *Notices of the AMS*, Volume 48, Number 9, October, pp. 992–996.

Rabin, M.O. (1959) Speed of computation and classification of recursive sets. In *Third Convention of the Scientific Society*, Israel, pp. 1–2.

Rabin, M.O. (1960) Degree of difficulty of computing a function and a partial ordering of recursive sets. Technical Report 1, O.N.R., Jerusalem.

Raisbeck, G. (1964) *Information Theory: An Introduction for Scientists and Engineers*. Cambridge: The MIT Press.

Ralston, A., Reilly, E.D. and Hemmendinger, D. (2000) *Encyclopedia of Computer Science*. London: Nature Publishing Group.

Randell, B. (1982) *The Origins of Digital Computers: Selected Papers*. New York: Springer-Verlag.

Randall, L. (2005) *Warped Passages*. New York: Ecco.

Reeves, A.H. (1965) The past, present, and future of pcm. *IEEE Spectrum*, Volume 2, May, pp. 58–63.

Reza, F.M. (1961) *An Introduction to Information Theory*. New York: McGraw-Hill Book Company.

Reza, F.M. (1994) *An Introduction to Information Theory*. New York: Dover.

Richards, R.K. (1955) *Arithmetic Operations in Digital Computers*. Princeton, NJ: D. Van Nostrand Company.

Riordan, M. and Hoddeson, L. (1997) *Crystal Fire: The Birth of the Information Age*. New York: W/W/ Norton and Company.

Rissanen, J. (1978) Modeling by the shortest data description. *Automatica*, Volume 14, pp. 465–471.

Rissanen, J. (2007) *Information and Complexity in Statistical Modeling*. New York: Springer.

Roszak, T. (1986) *The Cult of Information*. New York: Pantheon Books.

Ruelle, D. (1991) *Chance and Chaos*. Princeton: Princeton University Press.

Salomon, D. (2007) *Variable-Length Codes for Data Compression*. New York: Springer.

Sautoy, M. Du. (2006) Life begins at N=40. *New Scientist*, April 1, pp. 46–47.

Schillinger, J. (1976) *The Mathematical Basis of the Arts*. New York: Da Capo Press.

Schmalz, M.S. and Ritter, G.X. (2005) Computational complexity of object- based image compression. In *Mathematics of Data/Image Coding Compression and Encryption VIII, with Applications*, M.S. Schmalz (Ed.), *Proceedings of SPIE*, Volume 5915, pp. 82–91.

Schmidhuber, J. (2003) 35 year-old kind of science. *International Journal of High-Energy Physics*, Volume 43, Number 5, June. Based on a letter in the CERN Courier.

Schmidhuber, J. (2006) The computational universe. *American Scientist*, July–August, pp. 364–365.

Schnorr, C.P. (1973) Process complexity and effective random tests. *Journal of Computer Systems Science*, Volume 7, pp. 376–388.

Seibt, P. (2006) *Algorithmic Information Theory: Mathematics of Digital Information Processing*. New York: Springer-Verlag.

Seife, C. (2006) *Decoding the Universe*. New York: Viking Adult.

Selman, A.L. (1986) *Structure in Complexity Theory*. New York: Springer-Verlag.

Shankar, N. and van Rijsbergem, C.J. (1997) *Metamathematics, Machines and Godel's Proof*. Cambridge: Cambridge University Press.

Shannon, C.E. (1938) A symbolic analysis of relay and switching circuits. *Transactions of the American Institute of Electrical Engineers*, Volume 57, pp. 713–723. This is an abstract, same title, of Shannon's Master's of Science degree thesis awarded by the Massachusetts Institute of Technology.

Shannon, C.E. (1940) An algebra for theoretical genetics. Ph.D. Dissertation, Department of Mathematics, Massachusetts Institute of Technology, April 15, 69 pp.

Shannon, C.E. (1948) A mathematical theory of communication. *Bell System Technical Journal*, Volume 27, pp. 579–423 and 623–656.

Shannon, C.E. and Weaver, W. (1949) *The Mathematical Theory of Communication*. Urbana, IL: University of Illinois Press.

Shannon, C.E. (1950) A systematic notation for numbers. *American Mathematical Monthly*, Volume 57, pp. 90–93.

Shannon, C.E. (1956) The bandwagon. Editorial, *Institute of Radio Engineers, Transactions on Information Theory*, Volume IT-2, March, p. 3.

Shannon, C.E. (1956) A universal turing machine with two internal states. In *Automata Studies*, C.E. Shannon and J. McCarthy. Princeton: Princeton University Press.

Shannon, C.E. and McCarthy, J. (1956) *Automata Studies*. Princeton: Princeton University Press.

Shen, A. (1993) A strange application of Kolmogorov complexity. Centrum voor Wiskunde en Informatica, Amsterdam, the Netherlands, CS-R9328.

Shen, A., and Uspensky, V.A. (1993) Relations between varieties of Kolmogorov complexities. Centrum voor Wiskunde en Informatica, Amsterdam, the Netherlands, CS-R9329.

Siegfried, T. (2000) *The Bit and the Pendulum*. New York: John Wiley & Sons.

Simon, H. (1969) *The Science of the Artificial*. Cambridge: MIT Press.

Simpson, J.A. and Weiner, E.S.C. (1989a) *The Oxford English Dictionary*. Oxford: Clarendon Press. Volume II: B.B.C.-Chalypsography. 2nd Edition.

Simpson, J.A. and Weiner, E.S.C. (1989b) *The Oxford English Dictionary*. Oxford: Clarendon Press. Volume XII: Poise-Quelt. 2nd Edition.

Simpson, J.A. and Weiner, E.S.C. (1989c) *The Oxford English Dictionary*. Oxford: Clarendon Press. Volume XVIII: Thro-Unelucidated. 2nd Edition.

Singh, J. (1966) *Great Ideas in Information Theroy, Language and Cybernetics*. New York: Dover Publications.

Skyttner, L. (2001) *General Systems Theory*. Singapore: World Scientific Publishing Company.

Slepian, D. (1973) *Key Papers in the Development of Information Theory*. New York: IEEE Press.

Sloane, N.J.A. and Wyner, A.D. (1993) *Claude Elwood Shannon Collected Papers*. New York: IEEE Press.

Small, M. and Tse, C.K. (2002) Minimum description length neural networks for time series prediction. *Physical Review E*, Volume 66, Paper# 066701, pp. 1–12.

Smits, F.M. (Ed.) (1985) *A History of Engineering & Science in the Bell System: Electronics Technology, 1925–1975*. Murray Hill: AT&T Bell Laboratories.

Solana-Ortega, A. (2002) The information revolution is yet to come (an homage to Claude E. Shannon). *Bayesian Inference and Maximal Entropy Methods in Software Engineering*, Volume 617, May 14, pp. 458–473.

Solomonoff, R.J. (1960) A preliminary report on a general theory of inductive inference. Report ZTB-138, Zator Corporation, Cambridge, MA, November.

Solomonoff, R.J. (1964) A formal theory of inductive inference, Part I. *Information and Control*, Volume 7, Number 1, March, pp. 1–22.

Solomonoff, R.J. (1964a) A formal theory of inductive inference, Part II. *Information and Control*, Volume 7, Number 2, June, pp. 224–254.

Solomonoff, R.J. (1978) Complexity-based induction systems: Comparisons and convergence theorems. *IEEE Transactions on Information Theory*, Volume 24, pp. 422–432.

Solomonoff, R.J. (1995) The discovery of algorithmic probability: A guide for the programming of true creativity. In *Proceedings of Computational Learning Theory. Second European Conference, EuroCOLT'95*, Barcelona, Spain, March 13–15, P. Vitanyi (Ed.). New York: Springer-Verlag, pp. 1–22.

Solovay, R.M. (1975) Draft of a paper on Chaitin's work. Unpublished manuscript. IBM Thomas J. Watson Research Center, Yorktown Heights, New York, May.

Sommerer, C. and Mignonneau, L. (2003) Modeling complexity for interactive art works on the internet. In *Art and Complexity*, J. Casti and A. Karlqvist (Eds.). New York: Elsevier, pp. 85–107.

Sow, D.M. (2000) Algorithmic representation of visual information. Ph.D. Dissertation, awarded from Columbia University, New York.

Stein, D.L. (1989) *Lectures in the Sciences of Complexity*. Menlo Park: Addison-Wesley Publishing Company.

Stokes, J. H. (2006) AT&T labs vs. google labs: Not your grandfather's r&d. *ars technical*, July 29, pp. 1-6. Web site: http://arstechnica.com/news.ars/post/20060724-7340.html.

Stoll, C. (1995) *Silicon Snake Oil*. New York: Doubleday.

Svozil, K. (1996) Quantum algorithmic information theory. *Journal of Universal Computer Science*, Volume 2, Number 5, pp. 311–346.

Szilard, L. (1929) Uber die Entropieverminderung in einem thermodynamischen System bei eingriffen intelligenter Wesen. *Z. Physik.*, Volume 53, pp. 840–856. English translation 'On the decrease of entropy in a thermodynamic system by the intervention of intelligent beings', in Li and Vitanyi's *An Introduction to Kolmogorov Complexity and Its Applications*, 1997, p. 588.

Tadaki, K. (2008) A statistical mechanical interpretation of algorithmic information theory. arXiv: 0801.419v1, January 28.

Takahashi, H. (2004) Redundancy of universal coding, Kolmogorov complexity, and Hausdorff dimension. *IEEE Transactions on Information Theory*, Volume 50, Number 11, pp. 2727–2736.

Talbott, S.L. (1995) *The Future Does Not Compute*. Sebastopol, CA: O'Reilly & Associates.

Tapscott, D. (1998) *Growing Up Digital*. New York: McGraw-Hill.

Tarhal, B.M., Wolf, M.M. and Doherty, A.C. (2003) Quantum entanglement: A modern perspective. *Physics Today*, April, pp. 46–52.

Tenzeln, W. E. (1705) *Curieuse Bibliothec*. Frankfurt: Philipp Wilhelm Stock.

Thom, R. (1975) *Structural Stability and Morphogenesis*. Reading, MA: Addison-Wesley. Translation from the 1972 French edition.

Thompson, D.W. (1961) *On Growth and Form*. Cambridge: Cambridge University Press.

Tice, B.S. (2000) Physical laws as formal constraints to formal languages. Unpublished Manuscript.

Tice, B.S. (2003) *Two Models of Information*. Bloomington: 1st Books Publishers.

Tice, B.S. (2008) The use of a radix 5 base for transmission and storage of information. In *Integrated Optics: Devices, Materials, and Technologies XII, Proceedings of the SPIE*, Volume 6896, pp. 68961H-68961H-7.

Tran, N. (2007) The normalized compression distance and image distinguish ability. In *Human Vision and Electronic Imaging XII*, B.E. Rogowitz, T.N. Pappas and S.J. Daly (Eds.), *Proceedings of the SPIE*, Volume 6492, pp. 64921D.

Tukey, J.W. (1947) Sequential conversion of continuous data to digital data. Bell Laboratories Memorandum, September 1.

Uspensky, V.A. (1992) Complexity and entropy: An introduction to the theory of kolmogorov complexity. In *Kolmogorov Complexity and Computational Complexity*, O. Watanabe. Berlin: Springer-Verlag, pp. 85–102.

Verdu, S. (Ed.) (1999) *Information Theory: 50 Years of Discovery*. New York: Wiley-IEEE Press.

Vitanyi, P.M.B. (2001) Quantum Kolmogorov complexity based on classical descriptions. *IEEE Transactions on Information Theory*, Volume 47, Number 6, September, pp. 2464–2479.

Von Baeyer, H.C. (2004) *Information: The New Language of Science*. Cambridge: Harvard University Press.

Von Bertalanoffy, L. (1968) *General Systems Theory: Foundations, Development, Applications*. New York: George Braziller.

Von Neumann, J. and Morgenstern, G. (1944) *Theory of Games and Economic Behavior*. New York: John Wiley & Sons.

Wallace, C. and Boulton, D. (1968) An information measure for classification. *Computer Journal*, Volume 11, pp. 185–195.

Wallace, C. and Boulton. D. (1975) An invariant Bayes method for point estimation. *Classification Society Bulletin*, Volume 3, Number 3, pp. 11–34.

Wallace, C. and Freeman, P. (1987) Estimation and inference by compact coding. *Journal of the Royal Statistical Society*, Series B, Volume 49, pp. 240–251. Discussion: pp. 252–265.

Watanabe, O. (1992) *Kolmogorov Complexity and Computational Complexity*. Berlin: Springer-Verlag.

Weil, D. (2003) *Leaders of the Information Age*. New York: The H.W. Wilson Company.

Weinberg, G.M. (1975) *An Introduction to General Systems Thinking*. New York: John Wiley & Sons.

Weisstein, E.W. (2003) *CRC Concise Encyclopedia of Mathematics*. New York: Chapman & Hall/CRC.

Wexelblat, R.L. (1981) *History of Programming Languages*. New York: Academic Press. Chapter IV on 'LISP Session' by J. McCarthy, pp. 173–197.

Wheeler, J.A. (1994) *At Home in the Universe*. New York: AIP Press.

Wiener, N. (1948) *Cybernetics: Or Control and Communication in the Animal and the Machine*. New York: John Wiley & Sons.

Wiener, N. (1949) *Extrapolation, Interpolation, and Smoothing of Stationary Time Series, with Engineering Applications*. Cambridge: Technology Press of the Massachusetts Institute of Technology.

Wiener, N. (1950) *The Human Use of Human Beings*. Boston: Houghton Mifflin.

Wiener, N. (1953) *Ex-prodigy*. New York: Simon and Schuster.

Wiener, N. (1956) *I Am a Mathematician*. Garden City, NY: Doubleday.

Wiener, N. (1958) *Nonlinear Problems in Random Theory*. Cambridge, MA: The MIT Press.

Wiener, N. (1985) Wiener on cybernetics, information theory, and entropy. A paper by S. Watanabe in *Norbert Wiener: Collected Works with Commentaries*, P. Masani (Ed.), Cambridge: The MIT Press, pp. 215–218.

Wiener, N. (1985) A new concept of communication engineering. A 1949 paper by Norbert Wiener in *Norbert Wiener: Collected Works with Commentaries*, P. Masani (Ed.), Cambridge: The MIT Press, pp. 197–235.

Wikipedia (2006) Balanced ternary. Web address: http://en.wikipedia.org/wiki/Balanced_ternary.

Wikipedia (2006) Claude Elwood Shannon-Biography. Wikipedia Encyclopedia, 5 pp. Web address: http://en.wikipdeia.org/wiki/Claude-E-Shannon.

Wikipedia (2006) The Compendious Book on Calculation by Completion and Balancing. WikipediaEncyclopedia, 2 pp. Web address: http://en.wikipedia.org/wiki/The_Compendious_Book_on_Calculation_by_Completion_a...

Wikipedia (2006) Gregory Chaitin. Wikipedia Encyclopedia, 2 pp. Web address: http://en.wikipedia.org/wiki/Gregory-Chaitin.

Wikipedia (2006) Joseph Leo Doob. Wikipedia Encyclopedia, 3 pp. Web address: http://en.wikipedia.org/wiki/J.-L.-Doob

Wikipedia (2006) Ternary numeral system. Web address: http://en.wikipedia.org/wiki/Ternary_numeral_system.

Williams, M.R. (1997) *A History of Computing Technology*. Los Alamitos, CA: IEEE Computer Society Press.

Williams, T.I. (1978) *A History of Technology: Volume VII*. Oxford: Clarendon Press.

Winograd, J.M. and Nawab, S.H. (1995) Mixed-radix approach to incremental DFT refinement. *Proceedings of SPIE*, Volume 2563, pp. 418–429.

Wirth, N. (1976) *Algorithms + Data Structure = Programs*. Englewood Cliffs, NJ: Prentice-Hall.

Woesler, R. (2005) Problems of quantum theory may be solved by an emulation theory of quantum physics. *AIP Conference Proceedings*, Volume 750, February 15, pp. 378–381.

Wolfowitz, J. (1978) *Coding Theorems of Information Theory*. New York: Springer-Verlag.

Wolfram, S. (2002) *A New Kind of Science*. Champaign, IL: Wolfram Media.

Wright, R. (1988) *Three Scientists and Their Gods*. New York: Times Books.

Young, R.V. and Minderovic, Z. (1998) *Notable Mathematicians*. New York: Gale.

Yu, D., Liu, Y., Mu, R.Y. and Yang, S. (1998) Integrated system for image storage, retrieval, and transmission using wavelet transform. *Proceedings of SPIE*, Volume 3656, pp. 448–457.

Zayed, A.I. (1993) *Advances in Shannon's Sampling Theory*. London: CRC Press.

Zhao, J.H. (2003) Genetic number systems and haplotype analysis. *Computer Methods and Programs in Biomedicine*, Volume 70, Issue 1, January, pp. 1–9.

Zhao, J.H. and Sham, P.C. (1998) A method for calculating probability convolution using 'ternary' numbers with application in the determination of twin zygosity. *Computational Statistics and Data Analysis*, Volume 28, pp. 225–232.

Ziv, J. and Lempel, A. (1977) A universal algorithm for sequential data compression. *IEEE Transactions on Information Theory*, Volume 23, Number 3, May, pp. 337–343.

Zurek, W.H. (1991) *Complexity, Entropy and the Physics of Information*. Menlo Park: Addison-Wesley.

Zvonkin, A.K. and Levin, L.A. (1970) The complexity of finite objects and the development of the concepts of information and randomness by means of the theory of algorithms. *Russian Mathematical Surveys*, Volume 25, Number 6, pp, 82–124.

A note about the citations using Wikipedia. Wikipedia is a free on-line encyclopedia that is governed by voluntary entries by participants. One of the founders of Wikipedia, Larry Sanger, has found that both the participants and the quality of articles were wanting on Wikipedia (Poe, 2006: 93).[1] Please be aware that the validity of citations from Wikipedia may suffer accordingly.

---

[1] Poe, M. (2006) The hive. *The Atlantic Monthly*, September, 86–94.

Index

Note to reader: The Index references only main ideas and people cited in this work. A vast number of minor ideas and people can be found in the Review of Literature section of this book.

About the Author

Mr. Tice is the Director and CEO of Advanced Human Design that is located in Cupertino, California U.S.A. His research interests are in telecommunications, data compression, and linguistics. Mr. Tice is a member of the Association for Computing Machinery; ACM, the IEEE, SPIE, SIAM, American Mathematical Society; AMS, Sigma Xi, and is a Fellow of the Royal Statistical Society.

RIVER PUBLISHERS SERIES IN INFORMATION SCIENCE AND TECHNOLOGY

Other books in this series:

Volume 1
Traffic and Performance Engineering for Heterogeneous Networks
Demetres D. Kouvatsos
February 2009
ISBN 978-87-92329-16-5

Volume 2
Performance Modelling and Analysis of Heterogeneous Networks
Demetres D. Kouvatsos
February 2009
ISBN 978-87-92329-18-9

Volume 3
Mobility Management and Quality-of-Service for Heterogeneous Networks
Demetres D. Kouvatsos
March 2009
ISBN 978-87-92329-20-2